S0-BIL-771

Soil Fertility and Organic Matter as Critical Components of Production Systems

WITHDRAWN
NDSU

Soil Fertility and Organic Matter as Critical Components of Production Systems

Proceedings of a symposium sponsored by Division S-4, S-2, S-3, and S-8 of the Soil Science Society of America in Chicago, IL, 3 Dec. 1985.

Organizing Committee
R. F. Follett, chair
J. W. B. Stewart
C. V. Cole
J. F. Power

Editorial Committee
R. F. Follett, chair
J. W. B. Stewart
C. V. Cole

Editor-in-Chief SSSA
John J. Mortvedt

Editor-in-Chief ASA
D. R. Buxton

Managing Editor
Sherri H. Mickelson

SSSA Special Publication Number 19

**Soil Science Society of America, Inc.,
American Society of Agronomy, Inc., Publishers
Madison, WI USA**

1987

Cover Design: William R. Follett and Julia M. Whitty

Soil Science Society of America, Inc.
American Society of Agronomy, Inc.
677 South Segoe Road, Madison, WI 53711, USA

Copyright © 1987 by the Soil Science Society of America, Inc.
American Society of Agronomy, Inc.

ALL RIGHTS RESERVED UNDER THE U.S. COPYRIGHT
LAW OF 1978 (P.L. 94-553)

Any and all uses beyond the limitations of the "fair use" provision
of the law require written permission from the publisher(s) and/or
the author(s); not applicable to contributions prepared by officers
or employees of the U.S. Government as part of their official duties.

Second Printing 1989

Library of Congress Cataloging-in-Publication Data

Soil fertility and organic matter as critical
  components of production systems.

  (SSSA special publication ; no. 19)
  1. Soil management—Congresses. 2. Soil fertility—Congresses. 3.
Humus—Congresses. I. Follett, R. F. (Ronald F.), 1939–   . II. Soil
Science Society of America. Division S-4. III. American Society of
Agronomy. IV. Series.
S590.2.S625 1987   631.4′22   87-4574
ISBN 0-89118-782-0

Printed in the United States of America

Table of Contents

Foreword

This book is about the science of managing the soil and raising crops; specifically about the dynamics of organic matter in the soil and its role in soil fertility. It appears at an appropriate point in time. The chapters in this book develop the thread of information needed to understand the complex interactions between physical and biological soil factors, climatic factors, and farm management. Understanding the science of organic matter in the soil has progressed rapidly in recent years. There is a new appreciation for the essential presence of this soil component, there is a better understanding of the relationship between the organic and inorganic components of the soil production systems. Renewed interest comes from rapidly developing changes in soil management. Current farm management emphasizes economic yields. Pressures of pending environmental legislation are of concern as are the Soil and Water Conservation Provisions of the 1985 Farm Bill.

The American Society of Agronomy and the Soil Science Society of America are pleased to have sponsored the symposium "Soil Fertility and Organic Matter as Critical Components of Production Systems" at the 1985 Annual Meeting on 3 Dec. 1985 and this publication that resulted from it. The symposium was jointly sponsored by the Div. S-2, S-3, S-4, and S-8.

R. G. Gast, *president*
American Society of Agronomy

L. Boersma, *president*
Soil Science Society of America

Preface

This special publication represents the current thinking of scientists concerned with soil fertility and organic matter as critical components of production systems. Soil, climate, organic matter and nutrient requirements of crops affect soil management. Biological, physical, and chemical processes are involved in the balance between concurrent mineralization and immobilization, and nutrient availability. Thus, carbon inputs impact the supply of N, P, and S and the timing of their release in mineral form for both fertilized and unfertilized soils. Knowledge of these processes and their controls provides the basis for improved management practices to maintain or increase productivity of our soil resources. Because of the complexity of many of these relationships, improvements in soil fertility management and fertilizer recommendations will increasingly involve the use of computer models as part of the total technical support system.

This publication: (i) documents the important role of soil, climate, and management to the prediction of nutrient availability and use, (ii) describes controls on nutrient cycling and organic matter dynamics, and (iii) considers approaches for advisory services to use new technologies and to integrate information on organic matter dynamics and nutrient availability into models of crop production systems.

R. F. Follett, *editor*
USDA-ARS, Fort Collins, CO

J.W.B. Stewart, *editor*
Univ. of Saskatoon, Saskatchewan, Canada

C.V. Cole, *editor*
USDA-ARS, Fort Collins, CO

1 Integration of Organic Matter and Soil Fertility Concepts into Management Decisions[1]

J.W.B. Stewart, R.F. Follett, and C.V. Cole[2]

Production agriculture technology is making rapid advances toward integrating knowledge of organic matter and soil fertility management concepts along with other production factors. These production factors may include: tillage practices, previous crop in the rotation, use of organic wastes, and even the impact of pests and other yield-reducing factors. Recent articles such as Holt's (1984) thought-provoking paper dramatically demonstrate computers' potential to be integrated into nearly every aspect of a farm operation. Eventually, current local, regional, or national data bases will provide the farmer with information on fertilizer, seed, fuel and pesticide supplies and prices; weather; markets; and insect and disease predictions. Direct linkage of these data bases to on-farm computers will improve soil fertility management strategies, by designing crop production for specific soil and farming systems. The purpose of these strategies will include achievement of maximum economic yield, optimum fertilizer-use efficiency, soil improvement, and minimal degradation of the environment by fertilizer nutrients.

Computer simulation is a significant innovation and is expected to have a major impact on farming practices. The development of our knowledge of processes governing plant growth, soil organic matter formation, mineralization, and other production factors has progressed to the extent that they can be described mathematically in simulation models (Mackay and Barber, 1984; Silberbush and Barber, 1984). These models are interactive and controlled by environmental factors such as the availability of soil water, photosynthetic radiation, and climatic variables. Simulation of the processes involved can be linked to existing data bases to provide a powerful predictive capability to aid farm operators and managers in decision-making.

---

[1] Saskatchewan Institute of Pedology Pub. R478.

[2] Professor of soil science, Univ. of Saskatchewan, Saskatoon, SK Canada; soil scientists, USDA-ARS, Fort Collins, CO 80522, respectively.

Copyright © 1987 Soil Science Society of America and American Society of Agronomy, 677 S. Segoe Rd., Madison, WI 53711, USA. *Soil Fertility and Organic Matter as Critical Components of Production Systems,* SSSA Spec. Pub. no. 19.

CURRENT APPROACH TO SOIL TESTING

Soil fertility research and associated laboratory studies since the 1950s have established the efficacy of soil testing to predict nutrient needs of crops to be grown (Olson et al., 1982). This procedure is generally recognized as the best available method for diagnosing soil nutrient limitations before a crop is planted so that correction may be made in that year by appropriate fertilization. Most soil testing laboratories base recommendations on the calibration of chemical extractions of soils. Correlation studies must be conducted in various soil types to provide a basis for selection of laboratory tests that provide the best index of nutrient availability to plants (Hanway, 1973). As a result, innumerable methods have been proposed to measure the availability of nutrients such as P, N, K, and S in world soils (Olsen and Sommers, 1982; chapter 6 in this book). Correlation to plant yield establishes the validity of soil tests in the area where they are used. This type of correlation, however, is time-consuming and expensive. It will have to be repeated with changes in crop cultivars, different soil management practices, and different soil types. All methods depend upon the assessment of actual and potentially available nutrients in the soil prior to seeding.

Perhaps, as a consequence of the work involved in the correlative approach to soil testing, alternate methods of diagnosing nutrient deficiency have been developed. An example of a method is the maintenance concept, which implies that whatever the soil test level, a quantity of nutrient should be added to replace the amount the crop is expected to remove. The *conservation* of a soil's nutrient-supplying capacity has strong appeal but discounts the economic aspects to the farmer in cases where the soil's delivery capacity of a given nutrient may be adequate for top yields for several years (Olson et al., 1982). Careful comparison of the economic impact of soil test correlation and maintenance approaches to soil fertility clearly demonstrate that the maintenance concept is not economically viable in Nebraska soils. In addition, it was inferior to the nutrient-sufficiency approach to soil testing (Olson et al., 1982). Similar results have been obtained in other North American studies (Follett and Westfall, 1986).

Soil testing needs improvement with better sampling time to allow for seasonal changes in the amounts of available plant nutrients. Soil nitrate levels are highly variable as the net result of mineralization, immobilization, leaching, and denitrification with changes in soil water, soil temperature, and organic matter inputs (MacDuff and White, 1984, Fig. 1–1). Bicarbonate ($NaHCO_3$)-extractable P (Campbell et al., 1984; Jessop et al., 1977, Fig. 1–2 and 1–3) also fluctuates seasonally as a result of P-cycling processes (chapter 6 in this book), making accurate assessments difficult to obtain.

Natural systems have evolved in a manner that conserves nutrients. As an example, N is rarely lost in unfertilized grassland in large quantities by any process. Immobilization and mineralization processes are in syn-

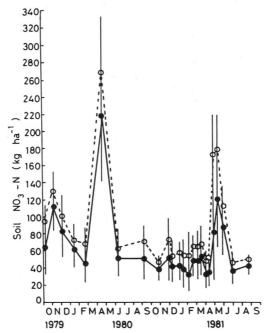

Fig. 1–1. Mean NO₃-N content in the top 1.0 m of soil between October 1979 and August 1981. (Vertical lines = 95% confidence limit of geometric mean; o = geometric mean; 0 = arithmetic mean.) From MacDuff and White (1984).

chrony with plant uptake that result in minimal nutrient losses. Crop production systems that mimic natural systems in terms of lack of soil disturbance and return of crop residues to the soil help maintain the synchrony of nutrient cycling processes and minimize nutrient and organic matter losses. In the past, soil degradation from poor management occurred even though short-term high crop production levels were obtained. With good management, sustained high levels of crop production provide excellent opportunities to significantly improve soil properties and their long-term productivity. High levels of crop production offer opportunity for the introduction of larger quantities of C into soils. Differences exist between plow tillage and no-tillage with regard to timing of C inputs. This can affect nutrient cycling and balance. Similarly, the quality of organic waste inputs (e.g., straw vs. legume vs. barnyard manure) affects nutrient composition of the soil solution and future nutrient-supplying power of soil.

Studies of soil organic matter and nutrient cycling (N, P, and S) emphasize the central role of C in nutrient cycling (Stevenson, 1986). In a system in dynamic equilibrium with interchanges governed by chemical, physical, and biological interactions, microbial activity is often depicted as a *wheel* rotating in the soil in response to energy (C) inputs and having a central role in nutrient transformations (chapter 6 in this book).

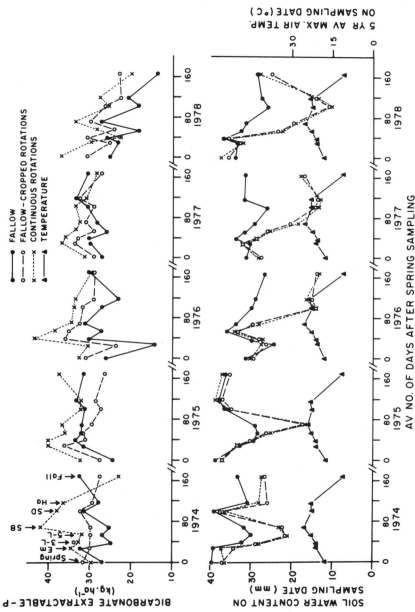

Fig. 1–2. Changes in bicarbonate-extractable P (NaHCO₃-P), soil water (0- to 150-mm depth) and 5-yr average maximum air temperatures on sampling dates for the period 1974 to 1978 in a crop rotation study in southwestern Saskatchewan. Data obtained from Campbell et al. (1984).

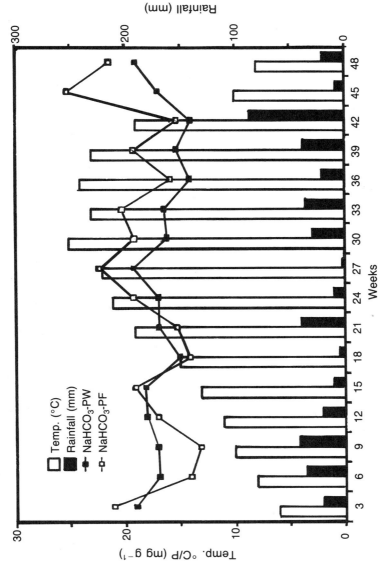

Fig. 1–3. Bicarbonate extractable P (NaHCO₃-P) levels for adjacent fallow and seeded wheat plots over a growing season. Adapted from Jessop et al. (1977).

Management often changes the timing of C inputs and thus, the management of C, in turn, can manipulate nutrient availability.

Follett et al (chapter 3 in this book) describes soil conservation practices that prevent erosion as well as available nutrient losses from topsoil. In summary, it can be stated that as management changes the timing of cultivation and organic matter incorporation, it will be necessary to reevaluate existing soil tests for applicability under the new management system. Under flexible cropping systems, this will involve costly field experimentation to provide adequate correlation data.

Attention should, therefore, focus on the necessity of understanding the processes involved in nutrient transformation in soil with emphasis on microbial activity. These processes have been studied in detail since 1975 (chapters 4, 5, 6, and 9 in this book). Computer simulation models are being developed to accurately depict the transformation of nutrients in soil. These models, in conjunction with reliable soil test values, can predict nutrient requirements to assist management decisions.

NEW APPROACHES TO SUPPLYING SOIL FERTILITY INFORMATION

Agriculture is increasingly utilizing computer capabilities for data capture and processing, automatic control, and aiding management decisions. The specific goals of obtaining maximum economic crop yields and optimum fertilizer-use efficiency while minimizing environmental degradation require accurate data on climate and reliable simulations of crop growth (including the soil environment for root growth and development), crop nutrient uptake, and soil nutrient status throughout the crop year. Shaffer and Clapp (1982), Baker et al. (1983), and Acock et al. (1985) describe the technology that already exists for computer prediction of profile root growth and activity for maize (*Zea mays* L.), cotton (*Gossypium hirsutum* L.), and soybean [*Glycine max* (L.) Merr.], respectively. Richardson and Wright (1984) and Williams and Renard (1985) describe models that predict daily precipitation, temperature, and solar radiation data for nearly anywhere in North America. The concept of irrigation scheduling using computer models and based upon climatic parameters is a widely accepted, if not standard, practice in most irrigated regions of developed and developing countries (Kruse, 1981).

The prediction of soil fertility relationships and improvements in fertilizer recommendations will increasingly involve the use of computer models that utilize data base and/or simulation approaches as described by Ward et al. and Cole et al. (chapters 8 and 9 in this book, respectively). We foresee an increased reliance on these tools as part of the total technical support system that Olson and Beaton (chapter 7 in this book) discuss. Availability of data and the degree of sophistication of other management predictions and recommendations will make it increasingly important to incorporate knowledge of chemical, physical, and biological processes into predictions of nutrient cycling and crop nutrient availability. Doran and Smith, McGill and Myers, and Stewart and Sharpley (chapters 4, 5, and 6 in this book, respectively) describe the type of processes involved and their relationship to nutrient cycling and availability. Conditions that may exist for various soil and climatic conditions (chapter 2 in this book) will influence the processes and their degree of importance to nutrient cycling and crop nutrient availability. Under certain conditions, for example, possibly for irrigated crops, nutrients can be scheduled along with irrigation water and the use efficiency of both water and nutrients optimized. Even in these irrigated systems, however, the influence of C inputs on the synchrony of nutrient mineralization and immobilization with crop nutrient requirements will be highly important to optimizing fertilizer-use efficiency. As more becomes known about the processes involved, the influence of key soil properties, and the influence of climate on these processes, accuracy of the simulations of crop nutrient availability for various soils and climatic conditions can be increasingly improved. The recalibrating of crop yield response to soil test for technological improvements such as tillage practice, crop variety, water management, or other improvements can be facilitated if the processes are adequately understood and properly simulated.

Agriculture is being increasingly held accountable for the environmental consequences resulting from agricultural activities. The impacts of fertilizer (or soil) management systems on either surface or groundwater or air quality as Follett et al. (chapter 3 in this book) describes are major types of environmental concern.

Finally, the economic impact and the computerization and availability of commodity prices must be recognized as having a major input into the decision-making processes. Certainly it is already in every farmer's mind as he plans his crop year, but—as with our assessment with nutrient inputs—the economic decision-making process is also becoming increasingly sophisticated. We anticipate there will be an integration of both agronomic and economic data into the decision-making process. This information will be available to the farmer using a desktop computer and will aid the entire decision-making process.

Before adequate economic recognition can be properly given to soil fertility and organic matter, however, there is a strong need to better define their impacts on crop yields. Relationships like those that Lucas et al. (1977) demonstrate and Cole et al. (chapter 9 in this book) discuss need to be incorporated into economic analyses. The availability of organic materials for inputs into agricultural production systems is part of the total management of nutrient resources (chapter 3 in this book). Differences in land and soil resources and the strong influence of climate on crop productivity (chapter 2 in this book) as they interact with the processes whereby soil fertility and organic matter affect crop yields will all be important to the overall economic recognition of the role of soil fertility and organic matter in production agriculture. Soil scientists need to better understand and adequately explain the economic assessment of the importance of soil organic matter and quantification of its importance to crop productivity.

REFERENCES

Acock, B., V.R. Reddy, F.D. Whisler, D.N. Baker, J.M. McKinion, H.F. Hodges, and K.F. Boote. 1985. The soybean crop simulator GLYCIM: Model documentation 1982. PB 85 171163/AS. U.S. Department of Agriculture, Washington, DC.

Baker, D.N., J.R. Lambert, and J.M. McKinion. 1983. GOSSYM: A simulation of cotton crop growth and yield. S.C. Agric. Exp. Stn. Tech. Bull. 1089.

Campbell, C.A., D.W. Read, G.E. Winklemann, and D.W. McAndrew. 1984. First twelve years of a long-term crop rotation study in southwestern Saskatchewan—bicarbonate-P distribution in soil and P uptake by the plants. Can. J. Soil Sci. 64: 125–137.

Follett, R.H., and D.G. Westfall. 1986. A procedure for conducting fertilizer recommendation comparison studies. J. Agron. Ed. 15: 27–29.

Hanway, J.J. 1973. Experimental methods for correlating and calibrating soil tests. p. 55–66. *In* L.M. Walsh and J.D. Beaton (ed.) Soil testing and plant analysis. Soil Science Society of America, Madison, WI.

Holt, D.A. 1984. Computers in production agriculture. Science 228: 422–427.

Jessop, R.S., B. Palmer, V.F. McClelland, and R. Jardine. 1977. Within-season variability of bicarbonate extractable phosphorus in wheat soils. Aust. J. Soil Res. 15: 167–170.

Kruse, E.G. (ed. com. chair.) 1981. Irrigation scheduling for water and energy conservation in the eighty's. ASAE Pub. 23–81.

Lucas, R.E., J.B. Holtman, and L.G. Connor. 1977. Soil carbon dynamics and cropping practices. p. 333–351. *In* W. Lockeretz (ed.) Agriculture and energy. Academic Press, New York.

MacDuff, J.H., and R.E. White. 1984. Components of the nitrogen cycle measured for crop and grassland soil-plant systems. Plant Soil 76: 35–47.

Mackay, A.D., and S.A. Barber. 1984. Soil temperature effects on root growth and phosphorus uptake by corn. Soil Sci. Soc. Am. J. 48: 818–823.

Olsen, S.R., and L.E. Sommers. 1982. Phosphorus. *In* A.L. Page et al. (ed.) Methods of soil analysis, Part 2, 2nd ed. Agronomy 9: 403–430.

Olson, R.A., K.D. Frank, P.H. Grabouski, and G.W. Rehm. 1982. Economic and agronomic impacts of varied philosophies of soil testing. Agron. J. 74: 492–499.

Richardson, C.W., and D.A. Wright. 1984. WGEN: A model for generating daily weather variables. USDA-ARS, ARS-8.

Shaffer, M.J., and C.E. Clapp. 1982. Root growth submodel. p. 149–173. *In* M.J. Shaffer and W.E. Larson (ed.) Nitrogen-tillage-residue management model—Technical Documentation. USDA-ARS, Washington, DC.

Silberbush, M., and S.A. Barber. 1984. Phosphorus and potassium uptake of field-grown soybean cultivars predicted by a simulation model. Soil Sci. Soc. Am. J. 48: 592–596.

Stevenson, F.J. 1986. Cycles of soil carbon, nitrogen, phosphorus, sulfur, micronutrients. John Wiley and Sons, New York.

Williams, J.R., and K.G. Renard. 1985. Assessment of soil erosion and crop productivity with process models (EPIC). p. 67–103. *In* R.F. Follett and B.A. Stewart (ed.) Soil erosion and crop productivity. American Society of Agronomy, Crop Science Society of America, Soil Science Society of America, Madison, WI.

2 Soil and Climate Effects upon Crop Productivity and Nutrient Use[1]

Richard W. Arnold and C. Allan Jones[2]

The effects of climate and soil on crop productivity are manifested at locations. When we stand on a spot of land and look around, our numerous sensors flood our personal data bases and cause us to select and group the information in meaningful patterns. Light and dark, hot and cold, near and far, brown and green, short and tall; our sensors gather a variety of data.

Weather patterns and climatic trends temper us as individuals. They also modify crop responses to agronomic practices. We cannot be certain, however, that a specific weather event will occur at a particular location, thereby assuring that risk is an ever-present element.

SOME CLIMATIC ATTRIBUTES OF LOCATION

We feel the air temperature and notice if the air is moving slowly or quickly; we sense the humidity in the air—ranging from very low to actual precipitation. The intensity and duration of solar radiation influence not only ourselves but also our surroundings. If land is flat for a long distance, climate tends to change slowly across space, but if the land is rolling or hilly, the micro-variations may modify the weather and climatic patterns in such a way that they will be reflected in the plants and animals that occur in those environments. Cool, damp air flows into low areas, and surface runoff water is redistributed throughout the landscape creating microclimates differing from the average conditions. Wind directions and intensities respond to landscape, and where snowfall is involved, the snow pack and melt waters modify the generalities that are used to characterize an area. Annual weather cycles impose temporal

[1] Contribution of USDA-SCS, Washington, DC and USDA-ARS, Temple, TX.

[2] Director, Soil Survey Division, USDA-SCS, P.O. Box 2890, USDA-SCS, Washington, DC 20013; and plant physiologist, USDA-ARS, Grassland, Soil, and Water Research Laboratory, P.O. Box 6112, Temple, TX 76503.

Copyright © 1987 Soil Science Society of America and American Society of Agronomy, 677 S. Segoe Rd., Madison, WI 53711, USA. *Soil Fertility and Organic Matter as Critical Components of Production Systems,* SSSA Spec. Pub. no. 19.

variabilty upon spatial variation. Thus, each spot on the earth's surface is unique and changes over time.

PEDOLOGIC ATTRIBUTES OF LOCATION

Beneath our feet is yet another universe whose behavior affects crop productivity. Soils have hundreds of properties that can be seen, felt, and measured. Sand, silt, and clay particles in combination are *texture;* small aggregates form *structural units* which have consistence and color. There are distributions of roots, stones, pores, organic matter, cations, anions, nutrients, and toxicants. Numerous soil properties combine and interact in fascinating ways to create soil qualities of great significance to land users.

Soil qualities such as water-holding capacity, available water capacity, favorable rooting volume, nutrient-supplying capacity, nutrient status, and erodibility result from soil properties and their interrelations. Most of the soil properties that affect these qualities exhibit spatial, rather than temporal, patterns. Combinations of soil qualities are commonly expressed as limitations for use or as suitabilities for specific purposes.

Many processes affect organic matter transformations and the dynamic cycling of nutrients essential for plant growth. Soil temperature and moisture regimes, that is, soil climate, are features that can be measured and characterized. Micro-differences observed aboveground also occur below the ground. Soil properties that affect these chemical and biological processes commonly exhibit temporal and spatial variability. Soil maps reveal some of the intricacies that exist in patterns bold enough and large enough for scientists to delineate.

Many of the observable soil features in landscapes are the result of thousands of years of development and modification. Thus, the relationships between soil characteristics and present-day climates are often indirect, and can be misleading.

GEOGRAPHIC COMBINATIONS OF CLIMATES AND SOILS

Climates, soils, and vegetation can be conceived as continua that vary over the earth's surface. It is common, however, to partition these continua into discrete groups whose members have similar properties. Most such classifications define the *central concept* of each group in order to facilitate communication and comparison. Thus, the groups can be described and mapped, and these maps often reveal broad correlations among climates, soils, and vegetation groups.

There have been numerous attempts to display soil data geographically. The UNESCO/FAO World Soil Map is one of the best known examples (FAO-UNESCO, 1970-1980). Many climatic classifications have also been attempted, as indicated at a UNESCO conference (Burgos,

1968) and a more recent one at the International Crops Research Institute for the Semi-Arid Tropics (ICRISAT, 1980). One of the better-known schemes for geographically displaying agro-climate is that of Food and Agriculture Organization (FAO, 1976, 1978). This scheme uses total radiation, daylength, temperature, rainfall, evapotranspiration, frost, and flooding as climatic parameters to assist in evaluating the suitability of a site for agriculture.

When soil and climatic data are simultaneously considered, agro-ecological zones can be defined and mapped. The FAO approach (FAO, 1978) has been primarily to outline areas having similar climates and then reduce predicted crop responses from a hypothetical high-level yield according to the limitations imposed by the soil conditions that exist in the zone. Social and economic conditions that appear reasonable at the time of analysis are also considered. This is a systematic and logical approach that has been useful in providing an overview of current situations and for projecting possible future consequences.

Another approach to defining agro-ecological zones is to combine soil associations into land units and to group land units into areas of similar soil and land conditions. The climates and farming systems within the delineated areas can then be described. An example of this approach is the description of the major land resource areas (MLRAs) of the USA (USDA, 1981a). The recent Resource Conservation Act analyses relied, in part, on information about the MLRAs and about selected soils within them (USDA, 1981b).

SOME CONCERNS

Several factors limit the use of these systems for planning purposes. Map scale and size of the delineated areas may be inappropriate. Accuracy of parameters and their interactions may not be well known, and definition of the land units can be no more reliable than the parameters used in their definition and delineation. For example, the probabilities of potential rainfall and the cropping systems that are economically viable under different scenarios are generally poorly documented (Samani and Hargreaves, 1985). As rainfall becomes more limited it often becomes less predictable, and the spector of risk looms larger.

A better understanding of weather events, their frequency, and their implications for crops, livestock, and human survival is needed to improve technology transfer and acceptance (Meyer, 1985). The models used to evaluate available options to the decision makers must consider the uncertainty of rainfall.

SITE POTENTIAL

The interactions of environmental factors at a site greatly influence the additions, transformations, translocations, and the losses of organic matter and plant nutrients. The dynamics of the physical, chemical, and biological systems affect various crop responses, including yield.

Environmental factors can be integrated in many ways to estimate the potential use of a site. One of the more familiar integrations is the FAO land evaluation scheme (FAO, 1984). Each crop has a unique set of responses to environmental properties such as energy (radiation), temperature, moisture, oxygen (soil drainage), nutrient availability, rooting conditions, soil crusting, air humidity, conditions for ripening the crop, flood hazard, climatic hazards, salinity, soil toxicities, and pest and disease pressures. Whether a site meets the requirements of a crop is a direct consequence of soil and climate properties.

Crops also have management requirements. Soil workability, mechanization potential, conditions for land preparation and clearance, conditions of storage and processing, timing of operations, access to the production unit, size of potential production units, and location can affect management requirements. Spatial soil patterns and temporal weather patterns affect most of these management qualities. Conservation requirements are affected by susceptibility to erosion and to soil degradation, both of which are associated with soil-climate interactions.

In the FAO system, these qualities of a site determine its potential for a specified use or crop. Expected crop yields are relative, for example, 100–80% yield, 80–60% yield, etc.; and they are based upon expert opinion and experimental evidence.

Simulation models provide another means to estimate site potential. Recently, general agricultural management simulation models have been developed. They can be used to evaluate site potential rapidly and cheaply for a wide variety of climatic conditions, soils, crops, and management alternatives. In addition to estimating crop yields, they can simulate the dynamic status of the soil environment and predict the effects of its degradation due to poor management.

SOME EXAMPLES

The erosion-productivity impact calculator (EPIC) (Williams and Renard, 1985) is a mathematical model that simulates soil erosion; how it affects soil properties; and the effects of these properties, weather, and management on crop yields. The EPIC was recently used to simulate the effects of tillage management on soil erosion and the effects of soil erosion on crop productivity in agro-ecological zones throughout the USA (USDA, 1981b). The zones were the 204 MLRAs that were reduced to 168 discrete areas for this study. Four tillage systems were simulated for typical crop rotations on typical soils in each area. Each 100-yr simulation used generated weather sequences typical of the area.

Though the simulations were used to estimate the long-term effects of soil erosion on crop productivity, they can also be used to estimate the effects of soil properties, weather, and management on site potential.

For this chapter, EPIC was used to simulate the effects of soil properties, weather, and management on crop yields (Tables 2-1 through 2-6).

Precipitation and energy available for evapotranspiration and growth largely determine potential rainfed crop production (FAO, 1984). These variables are also partially taken into account by the soil moisture and temperature regimes in the U.S. system of Soil Taxonomy. Their effects on site potential can be illustrated by comparing crop yields across transects of varying energy availability or precipitation.

The following effects may be partly masked or enhanced due to differences in the sets of soil properties that were available for making these

Table 2-1. Relationships between latitude and mean simulated winter wheat yields on soils in capability subclass 2E or 3E land for major land resource areas (MLRAs) with similar mean annual precipitation, and between 95° and 103° W Long.

MLRA	Soil series	Mean yield	Mean annual ppt	Mean harvest date	Latitude	County	State
		kg/ha	mm	day/month	degree		
83A	Knippa	1500	659	15 May	29.0	Atascosa	TX
78	Miles	1500	645	30 May	34.5	Harmon	OK
79	Shellabarger	2600	681	27 June	38.0	Stafford	KS
73	Harney	2500	600	29 June	39.2	Rooks	KS
102B	Moody	3000	651	16 July	42.5	Cedar	NE

Table 2-2. Relationship between mean annual precipitation and mean simulated maize† yields for major land resource areas (MLRAs) along latitude 40° N in the U.S. Corn Belt and Great Plains.

MLRA	Soil series	Mean yield	mean annual precipitation	Longitude	County	State
		kg/ha	mm	degrees		
72	Ulysses	2600	462	101.00	Thomas	KS
73	Uly	2800	597	99.25	Rooks	KS
74	Wells	4000	693	97.75	Ottowa	KS
106	Wymore	6200	784	96.00	Pawnee	NE
109	Adair	6300	939	93.00	Putman	MO
108	Tama	6500	941	90.50	McDonough	IL

†Spring tillage; maize, spring oats, legume hay rotation; soils in capability subclass 3E.

Table 2-3. Relationship between depth to root constraint and mean simulated soybean yield† for some soils in the southern Mississippi Valley Silty Uplands (MLRA 134).

Soil		Soil constraint		Mean yield
Series	Great group	Type	Depth	
			cm	kg/ha
Memphis	Hapludalfs	None	>200	2500
Collins	Udifluvents	Bulk density	152	2400
Loring	Fragiudalfs	Bulk density	86	1900
Calloway	Fragiudalfs	Bulk density	76	1800
Bonn	Natraqualfs	Sodium, bulk density	35	1700

†Spring tillage; cotton, soybean, oat, and wheat rotation.

Table 2–4. Relationship between depth of rooting and mean simulated maize and soybean yields† for several soils in the northern half of the Southern Piedmont (MLRA 136N).

Soil		Soil constraint		Mean yield	
Series	Suborder	Type	Depth	Maize	Soybean
			cm	———— kg/ha ————	
Cecil	Hapludults	Saprolite	122	6700	3000
Madison	Hapludults	Saprolite	89	5700	2600
Goldston	Dystrochrepts	Saprolite	53	3100	1300

†Spring tillage; maize, soybean, oat, wheat rotation.

Table 2–5. Effect of legumes on mean simulated annual fertilizer N requirements of entire crop rotations (not individual crops).

MLRA†	Soil	Rotations with two legumes			Rotations with one or no legumes		
		Crop‡	Yield	N	Crop	Yield	N
			——— kg/ha ———			——— kg/ha ———	
136N	Appling	M	6 400		M	6 200	
		S	2 900	37	S	2 800	94
		P	800		O	2 200	
					W	2 300	
116A	Peridge	M	5 200		M	5 000	
		O	2 400	43	B	2 400	137
		Hl	6 000		Hnl	6 000	
108	Tama	M	6 700		M	7 000	
		O	2 100	60	B	2 200	202
		Hl	10 000		Hnl	10 000	

†MLRA—major land resource area.
‡B—barley; M—maize; Hl—hay, legume; Hnl—hay, nonlegume; O—oat; P—peanut; S—soybean; W—wheat.

Table 2–6. Influence of topsoil organic carbon (OC) on annual fertilizer N requirements when yields for a given major land resource area (MLRA) are similar.

MLRA	Soil	OC	Rotation†	N
				kg/ha
53A, 52	Scobbe	1.30	Continuous W	27
	Williams	2.80	Continuous W	21
116B	Hobson	0.68	M-B-Hnl	146
	Baydo	1.68	M-B-Hnl	118
87	Silawa	0.65	B-W-Hnl	165
	Norwood	1.30	B-W-Hnl	137

†B—barley; M—maize; Hnl—hay, nonlegume; W—wheat.

comparisons. For example, soils in capability subclass 2E or 3E may differ in their texture, rooting depth, and nutrient-supplying capacity that influence crop productivity but are not criteria for erosion susceptibility.

In the central USA, energy available for evapotranspiration decreases with increasing latitude. Thus, moisture availability may differ even among areas with similar rainfall. For example, along a transect from 29N to 42N Lat, mean annual precipitation is similar, ranging from 600 to 681 mm. However, simulated winter wheat yields increased from about 1500 kg/ha at 29N Lat to about 3000 kg/ha near 42N Lat. A significant break appears to occur in the Great Plains between 35N and 38N Lat (Table 2-1). This break is associated with a cooler soil temperature regime (mesic vs. thermic) that, in turn, reflects a difference in energy available for evapotranspiration. Similar trends of crop yields for these same soils are found in the Soils-5 data base (Soil Interpretation Records, SCS-Soils-5, National Cooperative Soil Survey, USDA-SCS).

At similar latitudes in the central USA, annual precipitation is a major determinant of rainfed crop yields. For example, at 40N Lat comparisons of simulated maize (*Zea mays* L.) yields indicate a sigmoidal increase of yield with increasing precipitation within the range of 450 to 950 mm precipitation (Table 2-2). At the first location, the soil is in an aridic subgroup of an ustic soil moisture regime, the next two have ustic regimes, and the last three have udic soil moisture regimes. In this case as well, similar trends are found in the Soils-5 data base. These trends support the idea that energy and moisture are the driving forces that determine regions of potential crop production.

Within an agro-ecological zone, soil properties can limit crop production by affecting moisture and/or nutrient availability. For example, soil depth to root-restricting layers can affect crop yields (Table 2-3). In the southern Mississippi Valley Silty Uplands (MLRA 134), simulated soybean [*Glycine max* (L.) Merr.] yields of 1700 to 1800 kg/ha occur in soils with restrictive layers at depths of 35 to 75 cm. Yields increase to about 2400 kg/ha when the depth to a restrictive layer is at about 150 cm and then level off at about 2500 kg/ha at greater depths. Similar trends are reported for these soils in the Soils-5 data base.

In the Southern Piedmont (MLRA 136N) simulated yields of maize and soybean both increase markedly as soil depth increases from 50 cm to about 125 cm (Table 2-4). Soybean yields average 44% of the maize yields, which range from 3100 to 6700 kg/ha; however, yield responses are similar.

Location is significant to understand and select proper cropping systems and management practices. The impact of legumes in crop rotations is well known and is illustrated by estimating the amount of fertilizer N needed to sustain similar yields of nonlegumes during a 100-yr simulation (Table 2-5). The Appling soils in North Carolina are thermic Hapludults, the Peridge soils in Arkansas are warm mesic Paleudalfs, and the Tama soils in Iowa are cool mesic Argiudolls. In each area, the use of legumes in the crop rotation decreases the average annual amount of N fertilizer needed by the entire crop rotation (not individual crops) by 60 to 70%.

The influence of organic carbon on annual N fertilizer (Table 2-6) appears to be strongly associated with differences in soil temperature regimes. The Scobbe and Williams soils in South Dakota are frigid, the Hobson and Baydo soils in Missouri are mesic, and the Silawa and Norwood soils in Texas are thermic. In this example, each percent of additional organic carbon in the frigid soils is equivalent to more than 4 kg of fertilizer N/ha per year. The annual maintenance values for continuous wheat (*Triticum aestivum* L.) are also quite low (about 20 kg of N/ha). In the mesic soils, maintenance amounts of fertilizer N are much higher (about 120 kg of N/ha) and each percent of additional organic carbon is equivalent to about 30 kg of annual fertilizer N. In the thermic soils, maintenance levels are slightly more and each additional percent of organic carbon is equivalent to about 40 kg of fertilizer N annually.

CONCLUDING REMARKS

The preceding examples illustrate that soil and climate influence crop yields and fertilizer requirements. In some cases, solar radiation, temperature, or precipitation may limit crop growth. In other cases, soil conditions such as limited rooting depth, inadequate soil water and nutrient storage, or presence of toxic elements may restrict yields. Though we did not address the effects of improper soil, crop, or pest management in the previous examples, they can also reduce crop yields and damage the natural resource base. Land evaluation schemes take these factors into account when estimating site potential. Simulation models such as EPIC can also be used to evaluate such effects. In fact, simulation models can be used to evaluate previously untested combinations of soil, climate, and management, thereby reducing the amount of site-specific research needed to assess improved agricultural technology.

Simulation models can help improve awareness and understanding of agrotechnology transfer. However, models can only produce estimates as good as the soil, climate, and management data used to run them. It has not been possible to limit differences of soil properties to the extent the authors hoped for. The authors believe that the trends obtained are reasonable estimates of effects. Close attention should be given to developing data bases to supply this information. Only when accurate models are used with accurate data can decision makers be supplied with tools to confront changing physical, biological, social, and economic conditions.

REFERENCES

Burgos, J.J. 1968. World trends in agroclimatic surveys. p. 211–224. *In* Agroclimatological methods: Proceedings of the Reading Symposium, UNESCO, UK.

Food and Agriculture Organization. 1976. A framework for land evaluation. Soils Bull. 32, FAO, Rome.

-----. 1978. Report on the agro-ecological zones project, Vol. 1. Methodology and results for Africa. World Soil Resources Rep. 48/1. FAO, Rome.

-----. 1984. Guidelines: land evaluation for rainfed agriculture. Soils Bull. 52, FAO, Rome.

FAO-UNESCO. 1970–1980; Soil map of the world. FAO of the United Nations, Rome.

International Crops Research Institute for the Semi-Arid Tropics (ICRISAT). 1980. Climatic classification: a consultant's meeting, 14-16 April. ICRISAT Center, Patancheru, India.

Meyer, R.E. 1985. Rainfed agricultural research network in Africa. Draft working paper, mimeo. S&T/RNE/OA/USAID, Washington, DC.

Samani, Z.A., and G.H. Hargreaves. 1985. A crop water evaluation manual for Brazil. International Irrigation Center. Utah State University, Logan.

U.S. Department of Agriculture. 1981a. Land resource regions and major land resource areas of the United States. USDA-SCS Agric. Handb. 296. U.S. Government Printing Office, Washington, DC.

————. 1981b. Soil, water, and related resources in the United States: analysis of resource trends. 1980 RCA Appraisal, Part II. USDA, Washington, DC.

Williams, J.R., and K.G. Renard. 1985. Assessments of soil erosion and crop productivity with process models (EPIC). p. 67–103. *In* R.F. Follett and B.A. Stewart (ed.) Soil erosion and crop productivity. American Society of Agronomy, Crop Science Society of America, and Soil Science Society of America, Madison, WI.

3   Conservation Practices: Relation to the Management of Plant Nutrients for Crop Production[1]

R. F. Follett, S. C. Gupta, and P. G. Hunt[2]

Increased use of conservation practices, especially conservation tillage, helps sustain or enhance the productivity of soil resources by reducing soil erosion losses of plant nutrients and soil organic matter. This change in cropland management, from almost complete turning of the surface soil, has occurred primarily since 1970 in the USA. The change has been encouraged by a national concern for erosion control and maintenance of soil productivity, but offers opportunities for improved farm profitability through reduced resource input. This chapter focuses on the USA because of the accessibility of information. The topics discussed, however, are also pertinent to other countries.

The type of crop production system selected can especially influence soil fertility and organic matter because of effects on the soil's biological, chemical, and physical components. *Soil fertility* refers to the capability of the soil to supply nutrients that enhance plant growth. *Soil productivity* is the soil's ability to produce a crop. Productivity is a function of a soil's natural fertility plus nutrients added as fertilizer, organic residues, and other sources; soil physical and biological properties; climate; management; and other non-inherent factors used to produce crops (Follett and Wilkinson, 1985). Soil organic matter concentration is a critical component of soil productivity that can be changed by altering cropland management practices. This is important because organic matter improves soil-fertility, -tilth, and -erosion control; water-infiltration and storage; and the soil's ability to bind and promote microbial breakdown of toxic substances. The dynamic effect of soil organic matter suggests that perhaps the best opportunity for sustaining or enhancing the long-term fertility and productivity of our cropland soils can be achieved by

[1] Contribution from the Soil-Plant-Nutrient Research Unit, USDA-ARS, Fort Collins, CO and Coastal Plains Soil and Water Conservation Research Center, Florence, SC 29501, in cooperation with the Dep. of Soil Science, Univ. of Minnesota, St. Paul, MN.

[2] Soil scientist, USDA-ARS, P.O. Box E, Fort Collins, CO 80522; Associate professor of soil science, Univ. of Minnesota, St. Paul, MN 55108; and soil scientist, USDA-ARS, P.O. Box 3039, Florence, SC 29501, respectively.

Copyright © 1987 Soil Science Society of America and American Society of Agronomy, 677 S. Segoe Rd., Madison, WI 53711, USA. *Soil Fertility and Organic Matter as Critical Components of Production Systems,* SSSA Spec. Pub. no. 19.

improving management of this soil component. This can be accomplished by developing improved conservation tillage practices that effectively utilize crop residues, as well as other added organic and/or inorganic crop production components.

The objectives of this chapter are to: (i) provide an overview of the use of conservation tillage in the USA; (ii) review plant nutrient sources (i.e., fertilizer, organic residues, and symbiotic N_2 fixation); (iii) discuss the ways in which conservation tillage and emerging cropping systems need to influence plant nutrient management; and (iv) discuss current and future issues relative to assessing the effectiveness of soil conservation practices for their effects on soil fertility.

CONSERVATION TILLAGE IN THE USA

Conservation tillage is generally an umbrella term to describe tillage practices that conserve soil and water (Mannering and Fenster, 1983). The Soil Conservation Society of America's (1982) *Resource Conservation Glossary* defines conservation tillage as "any tillage system that reduces loss of soil or water relative to conventional tillage; often a form of non-inversion tillage that returns protective amounts of residue mulch on the surface." *No-Till Farmer* magazine (Christensen and Magleby, 1983) uses the following definitions.

1. *Conventional tillage*—Where 100% of the topsoil is mixed or inverted by plowing, power tillering, or multiple disking.
2. *Minimum tillage*—Limited tillage, but where the total field surface is worked by tillage equipment.
3. *No-till*—Only the intermediate seed zone is prepared. Approximately 25% of the surface area could be worked. It could be no-till, till plant, chisel plant, or rotary-strip plant. It includes forms of conservation and mulch tillage.

Recently, Magleby et al. (1985) reported that The Conservation Tillage Information Center (CTIC), *No-Till Farmer* magazine, The National Resources Inventory (NRI), and Farm Production Expenditure Survey (FPES) estimated the amount of land under conservation tillage was 35.1, 36.9, 34.3, and 24.5 million ha; respectively. Conservation tillage includes the minimum and no-till categories as defined by *No-Till Farmer*. By contrast, CTIC uses a definition of conservation tillage to include tillage planting systems in which 30% or more of the surface is covered with residue just after planting (Conservation Tillage Information Center, 1985). Since each of the above sources uses a different information base for arriving at its estimate and the NRI estimate is for 1982, it is not surprising that their estimates differ. Yet, three of the sources were generally similar. Estimates by FPES may have been low because of the procedures used or as a result of not accounting for conservation tillage applied to planted acreage diverted to U.S. government programs. Since the *No-Till Farmer* magazine (Lessiter, 1974, 1975, 1977, 1979, 1981, 1983, 1985) provides annual data on changes in tillage practices nationally for the last several years, their data and definitions are most useful for the purposes of this chapter.

Table 3–1 shows the changes in the use of various tillage systems nationally from 1973 through 1985 (estimated). During this 13-yr period, the percentage of the total tilled area devoted to conservation tillage (minimum plus no-till) has nearly doubled (from 18 to 35%) while the percentage in conventional tillage has decreased from 82 to 65%.

Figure 3–1 shows that the use of conservation tillage varies widely among the 10 farm production regions of the USA as adapted from Christensen and Magleby (1983) for 1981. We chose 1981 for our illustration because it was a stable crop production year and the data avoids the impacts on land use patterns due to recent U.S. government programs such as Payment in Kind (PIK), (USDA, 1984) and recent severe financial distress of U.S. farmers (USDA, 1985a). Percentage of cropland area in conservation tillage (minimum plus no-till) is highest in the Southeast, Northeast, and Appalachian regions. The Corn Belt, Northern Plains, and Mountain regions all have about one third of their cropland area under conservation tillage. The largest areas in conservation tillage are in the Corn Belt (11.6 million ha) and Northern Plains (10.7 million ha) (Fig. 3–1).

TILLAGE MANAGEMENT REGIONS

The amount of conservation among the various crop production regions depends on a number of factors including soil types, climate, crops, and general cropping practices. Because of the importance of these factors to the adoption of conservation tillage, Allmaras et al. (1985)

Table 3-1. Various tillage systems used in the continental USA from 1973 to 1985.

Year	Total hectares tilled†		
	Minimum	No-till	Conventional
	%		
1973	15.8	2.0	82.2
1974	17.0	2.1	80.9
1975	17.6	2.4	80.0
1976	18.4	2.7	78.9
1977	21.0	2.5	76.5
1978	22.7	2.4	74.9
1979	23.9	2.5	73.6
1980	27.5	2.4	70.1
1981	29.1	2.9	68.0
1982	31.7	3.7	64.6
1983	35.0	4.0	61.0
1984	25.8	4.5	69.7
1985 (estimate)	29.9	5.0	65.1

†Based upon summation of acres in minimum, no-till, and conventional practices as reported in *No-Till Farmer* (Lessiter, 1974, 1975, 1977, 1979, 1981, 1983, 1985).

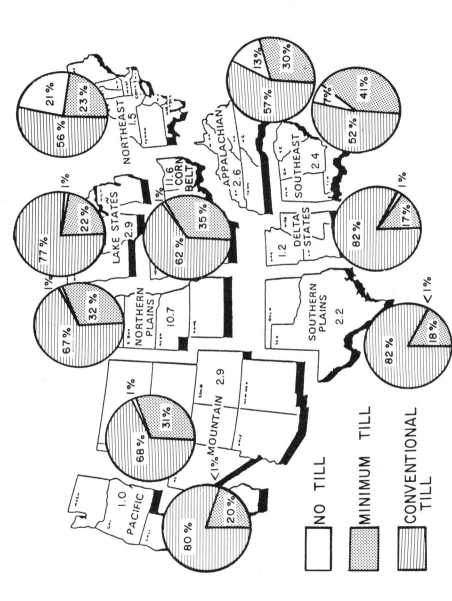

Fig. 3–1. Percentages of land in various tillage and millions of hectares of conservation tillage (minimum tillage plus no-till) by farm production region in 1981. Adapted from Christensen and Magleby (1983).

recently identified seven Tillage Management Regions (TMRs) across the USA (Fig. 3–2).

Their identification of land areas into TMRs corresponded to land resource regions (LRRs) or in some instances into major land resource areas (MLRAs) (USDA, 1981b). Within a TMR, soil type, local weather conditions, and crop rotations will vary. In addition, primarily rainfed agriculture is considered since irrigated agriculture may have a different sensitivity to the environment in the TMR. The bases on which the TMRs were developed include conservation and tillage needs and functions of conservation tillage. For example, depending upon soil types, topographic characteristics, and other factors, improved water conservation and use and water erosion control are major conservation needs in all of the TMRs. Wind erosion control is a major conservation need in the Northern and Southern Plains, Pacific Northwest, and Corn Belt. Soil temperature management is important in the Corn Belt and the Northern Great Plains to overcome cold soil conditions. In addition, soil drainage is a major need in the Corn Belt and management of restrictive soil layers is important in the Coastal Plains and Corn Belt TMRs (Allmaras et al., 1985). All of these needs can be influenced by the type of tillage management being used. Expected conservation and tillage needs are related to the climate and principal crops shown in Table 3–2.

Production agronomists have several management options for developing successful soil-crop managment systems. These options include tillage, residue placement, fertilization, crop rotation, and pest control practices. One change in a component of the integrated system may sig-

Fig. 3–2. Tillage management regions (TMRs) in the conterminious USA (Allmaras et al., 1985).

Table 3–2. Climatic characteristics, major crops, and typical major land resource areas (MLRAs) (USDA, 1981b) in seven tillage management regions (TMRs) of the USA (Allmaras et al., 1985).

Tillage management region	Typical MLRAs†	Mean monthly temperature‡		Mean annual pan evaporation‡	Mean annual precipitation‡	Frost-free season‡	Growing season annual precipitation	Major crops
		January	July					
		°C		cm		days	% of annual	
Corn Belt	103,104,106,107 108,110,111,114	−5 −10 to 0	22 20 to 25	80 70 to 100	80 50 to 100	180 130 to 230	50	Maize, soybean, wheat, hay, feed grains
Eastern Uplands	121,134,147	3 0 to 10	22 20 to 27	100 70 to 100	120 90 to 150	190 120 to 280	<50 in south >50 in north	Cotton, maize, soybean, small grain
Piedmont	136,148	5 0 to 10	24 20 to 25	90 80 to 100	120 90 to 150	200 160 to 240	<60	Maize, wheat, grain, sorghum, soybean, forages, cotton
Coastal Plains	133A	8 5 to 10	25 25 to 27	105 100 to 120	130 100 to 155	240 200 to 280	>60 in east <60 in west	Maize, soybean, grain, sorghum, small grains, cotton

(continued on next page)

Table 3–2. Continued.

Tillage management region	Typical MLRAs†	Mean monthly temperature‡ January	Mean monthly temperature‡ July	Mean annual pan evaporation‡	Mean annual precipitation‡	Frost-free season‡	Growing season annual precipitation	Major crops
		°C	°C	cm	cm	days	% of annual	
Southern Great Plains	73,77,78	5 / −3 to 10	27 / 26 to 28	150 / 120 to 190	60 / 40 to 100	170 / 130 to 250	>70	Wheat, maize, soybean, grain, sorghum, cotton, forages.
Northern Great Plains	55B,56,72,75	−10 / −15 to 3	22 / 20 to 25	90 / 70 to 130	40 / 25 to 60	130 / 100 to 160	>70	Wheat, maize, soybean, grain, sorghum, sunflower, feed grains, forages
Pacific Northwest	7,8,9	0 / −5 to 0	18 / 15 to 20	90 / 80 to 100	35 / 15 to 60	160 / 100 to 200	<45	Wheat, peas, barley, lentils

†Major land resources areas are (7) Columbia Basin, (8) Columbia Plateau, (9) Palouse Nez-Perce Prairie, (55B) Central Black Glaciated Plains, (56) Red River Valley, (72) Central High Tableland, (73) Rolling Plains and Breaks, (75) Central Loess Plains, (77) Southern High Plains, (78) Central Rolling Plains, (103) Central Iowa and Minnesota Till Prairies, (104) Eastern Iowa and Minnesota Till Prairies, (106) Nebraska and Kansas Loess-Drift Hills, (107) Iowa and Missouri Deep Loess Hills, (108) Illinois and Iowa Deep Loess Drift, (110) Northern Illinois and Indiana Heavy Till Plain, (111) Indiana and Ohio Till Plain, (114) Southern Illinois and Indiana Thin Loess and Till Plain, (121) Kentucky Bluegrass, (133A) Southern Coastal Plains, (134) Southern Mississippi Valley Silty Uplands, (136) Southern Piedmont, (147) Northern Appalachian Ridges and Valleys, and (148) Northern Piedmont.

‡These observations are arranged to show characteristic value on first line and range on second line.

nificantly impact the performance of the entire system to accomplish certain functions for which it is intended.

For example, conservation tillage systems can perform certain functions as Follett and Bauer (1986) describe including: (i) controlling rill and inter-rill erosion, (ii) conveying runoff water non-erosively, (iii) preventing wind erosion, (iv) protecting soil fertility, (v) maintaining soil organic matter, (vi) enhancing root-zone characteristics for plant growth, (vii) improving water infiltration, (viii) soil temperature management, and (ix) possibly others such as pest control. Addressing the major conservation needs of the various TMRs requires that many of the above functions be accomplished simultaneously.

Considerable progress has been made concerning the development of conservation and tillage systems to accomplish functions associated with soil physical components. In the future, increased emphasis is needed to identify and accomplish functions associated with the protection and improvement of the soil fertility component of soil productivity.

NUTRIENT RESOURCES

Figure 3-3 schematically presents nutrient sources in crop production systems and the components of plant nutrient cycling that conservation practices influence. In Figure 3-3, the plant nutrients that are influenced by conservation practices are those already present in the soil and those added or returned to the soil at a time when the conservation

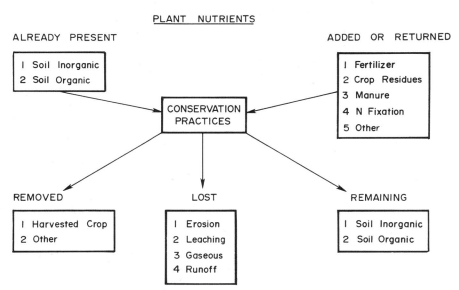

Fig. 3-3. Conceptual role of conservation practices in the management of plant nutrients and their eventual redistribution and fate.

practice(s) can influence what happens to them. To the degree that there is an effect by conservation practice, it is on those plant nutrients removed in harvested crops, lost to the environment, and/or remaining in the soil at the end of a given time period.

The relationship in Fig. 3–3 that shows the effect of conservation practices on soil fertility can be described as a nutrient budget in Eq. [1], as follows:

$$RN_{tn} = \sum_{t}^{tn}(AP_t + AR_{\Delta t} - RM_{\Delta t} - L_{\Delta t} \qquad [1]$$

where

RN = soil inorganic and organic nutrients remaining at time (tn),
AP = soil inorganic and organic nutrients present at time t,
AR = inorganic and organic nutrients added or returned to the soil during the time interval Δt,
RM = plant nutrients removed with the harvested product during the time interval Δt,
L = inorganic and organic nutrients lost during the time interval Δt,
t = the beginning time for imposing and determining the effectiveness of the conservation practice(s) used for conserving plant nutrients,
tn = the ending time for determining the effectiveness of the conservation practices used for conserving plant nutrients, and
Δt = The time interval between t and tn.

If $RN_{tn} \geqq AP_t$, then the reservoir of plant nutrients in the soil should be maintained or increased by the conservation practices being used. Useful indices such as increased organic carbon or organic nutrient (nutrients in soil organic matter) content may reflect this relationship. In addition, some interpretation might be made of the degree to which soil productivity is being maintained or even enhanced. Even if $RN_{tn} < AP_t$, it may not be of major concern depending upon their rates of change. Also, if $RM_{\Delta t}$ is larger and $L_{\Delta t}$ is smaller than they were before the use of conservation practices was begun, then removal of harvested crop yields would be larger while losses from the soil-plant system would be smaller because of the conservation practices. The most desirable combination of relationships is for $RN_{tn} \geqq AP_t$, while $RM_{\Delta t}$ increases and $L_{\Delta t}$ decreases.

Other relationships can be developed from Eq. [1] and the concepts shown in Fig.3–3. Irrespective, it is increasingly important to begin establishing the functions desired of conservation practices for cropland and the criteria whereby the effectiveness of the conservation practices can be measured, evaluated, and hopefully improved.

The discussion that follows will present quantitative information from the USA for some parts of Fig. 3–3. In general, quantitative information is given where we are reasonably confident that: (i) the methods used to arrive at those data can be substantiated (i.e., nutrients added or returned in fertilizer and crop residues), (ii) the relative values provide a perspective on how conservation practices influence the management of nutrient resources (e.g., controlling soil erosion), and (iii) the values given will not be misleading.

Nutrients Already Present in Cropland

Generally, there is a lack of any coherent soil test data base that can be used to assess inherent soil fertility and soil nutrient status on a nationwide basis. Existing data bases are often presented as a percent of samples in high, medium, and low fertility status. Thus, we did not try to estimate total quantities of inorganic- and organic-soil nutrients found on cropland on either a national or regional basis. Irrespective, the nutrient quantities that exist are enormous, but not inexhaustible. They are, perhaps, among any nation's most valuable resources and serve as a vast reservoir through which both inorganic and organic nutrients are added and/or returned to cropland soils.

Soil supplies 13 of the 16 elements that are known to be essential for crop growth of which N, P, and K are most commonly deficient in agricultural soils. Secondary- and micro-nutrient deficiencies have been widely documented in some soils, with S, Zn, and B being the most common. In order to maintain high crop yields, the addition and release of nutrients, particularly N, P, and K must be in balance so that the nutrients are always at a level of availability to attain economic (preferably maximum) yields.

With plowing and secondary tillage operations, the rate of decay of soil organic matter and release of its associated nutrients is related to the proportion of old and new humus, aeration, moisture, and temperature (Lucas et al., 1977). Reduced tillage and especially no till, when used continuously for several years, can result in soils that have higher organic matter than those that are plowed (Blevins et al., 1977; Lamb et al., 1985; Stanford et al., 1973). Additions of organic residues are a major factor in maintaining or increasing soil organic matter (Power and Legg, 1978). Larson et al. (1972, 1978) demonstrated that, after 11 yr, soil organic carbon content was a linear function of the amount of crop residue added (Fig. 3-4) in Iowa. It was estimated that about 5 Mg ha^{-1} of crop residues were needed to maintain the original C content in the soil under conventional (plowed) tillage conditions. Rasmussen (1980) reported the same observation and estimate in Oregon, based upon a 45-yr experiment sampled at 11-yr intervals. It is likely that under conservation tillage conditions, lower amounts of crop residues would be required to maintain the original C content in the soil. Addition of N fertilizer and consequent higher dry matter production also help increase and maintain higher levels of soil organic matter (Blevins et al., 1983; Meisinger et al., 1985).

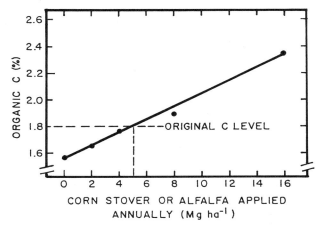

Fig. 3–4. Carbon content of a Typic Hapludoll as influenced by amounts of corn stover or alfalfa hay added to the soil for each of 11 consecutive yr. Soil was cropped to maize with conventional tillage (Larson et al., 1978).

Nutrients Added or Returned to Cropland

Fertilizer

Fertilizer is defined as *"any organic or inorganic material of natural or synthetic origin which is added to a soil to supply certain elements essential to the growth of plants"* (Soil Conservation Society of America, 1982). For purposes of this discussion, inorganic materials or commercial fertilizers are referred to in this chapter. The use of such fertilizers is now an economic necessity on most cropland soils. For example, yield increases of maize (*Zea mays* L.) attributed to increases in fertilizer use range from 20 to 50% (Walsh, 1985). Fertilizers represent the major input of added nutrients to croplands. Addition of fertilizers to the soil result in increased soil solution nutrient concentrations at the point of application; therefore those nutrients are highly available for plant uptake. In addition, commercial fertilizers are the most controllable source of nutrients for crop production. Through the use of appropriate rates, placement, sources, and application times of fertilizer, it is possible to supply nutrients reasonably close to economically optimum levels. In contrast, it is difficult to fine-tune the amount or timing of soil organic or inorganic nutrient release to optimize availability for crop uptake.

When the cost of applied fertilizer is low relative to the value of the crops, there is strong incentive to avoid any deficiencies and the use of larger fertilizer amounts per unit of yield response are made than for crops of lower relative value. Thus, certain crops, such as vegetables, may often be over-fertilized and inefficient use of fertilizers may result. Furthermore, differences exist in the ability of different crops to absorb nutrients from the soil. Plants vary greatly in their ability to take up applied fertilizers (especially N). For example, N fertilizer uptake efficiency for lettuce (*Lactuca sativa* L.) may only be 12 to 25% while for maize it may be close to 50% (Broadbent, 1985). Among the plant characteristics influencing nutrient uptake are nature and extent of root system, rate of crop growth, nutrient requirements during the growing season, and duration of crop growth.

A good soil-testing program is essential to sound fertilizer use. The soil test value is the starting point. Soil test is a means to evaluate the ability of the soil to supply these nutrients. Soil tests also evaluate carryover levels of past fertilizer programs. More carryover can be expected with high application rates and following droughty years. Nutrient carryover from manured soils and from the return of crop residues may also occur since not all of the nutrients are released during the 1st yr after application.

Figure 3–5 shows the increase in the use of commercial fertilizers for supplying N, P, and K to crops in the USA since 1955 (Hargett and Berry, 1985; USDA, 1981a, 1985b). Commercial fertilizer use on cropland began to increase most rapidly in the period following the end of World War II. In 1954, about 30% of all harvested crops and cropland pasture in the USA received fertilizer (Adams et al., 1958). In 1947 and 1954, estimates were that 44 and 60% of the harvested cropland planted to maize and 18 and 28% of the harvested cropland planted to wheat (*Triticum aestivum* L.) were fertilized, respectively. Average N, P, and K rates of fertilization for maize in 1954 were 30, 14, and 23 kg ha^{-1} and for wheat were 30, 13, and 18 kg ha^{-1} (Adams et al., 1958) In 1984, 97 and 76% of the harvested cropland planted to maize or wheat received

PLANT NUTRIENT CONSUMPTION

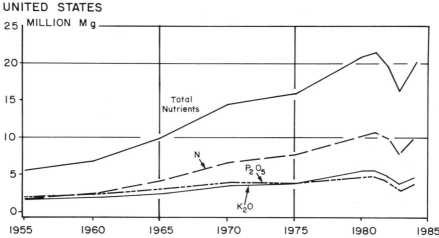

Fig. 3–5. Plant nutrient consumption in the USA between 1955 and 1984 (Hargett and Berry, 1985; USDA, 1981a, 1985a).

fertilizer, respectively. Average N, P, and K rates of fertilization for maize in 1984 are estimated to be 155, 32, and 81 kg ha^{-1} and for wheat 70, 18, and 43 ka ha^{-1}, (USDA, 1985b). Table 3–3 shows the 1977 use of fertilizer N, P, and K by crop production region (USDA, 1981a). The year 1977 was chosen to more nearly correspond with the data for manure and crop residues shown in Tables 3–4 and 5, as well as to avoid the period of economic distress for U.S. farmers that has occurred since 1982 (USDA, 1985a, 1985b).

Organic Residues

Organic residues are a tremendous natural resource for providing plant nutrients and C to help maintain soil fertility, organic matter, and tilth of soils. Organic residue, as well as fertilizers, are important nutrient sources. This discussion will address primarily N, P, and K. The organic residues available in the USA for use on soils include livestock wastes, crop residues, sewage sludge and septage, food processing wastes, industrial organic wastes, logging and wood manufacturing wastes, and municipal refuse. Estimates are that, of total annual production of organic wastes, animal manure and crop residues account for about 22 and 54%, respectively (USDA, 1978). All other organic waste sources, in terms of current land use and probability of increased use, account for < 1% of the total organic waste production in the USA or else they have a low to very low probability for increased use on agricultural lands (USDA, 1978). Therefore, this part of the discussion will deal only with animal manure and crop residues.

Animal Manure—Manure from animals represents about 22% of all organic wastes produced in the USA and refers to feces and urine excreted by dairy or beef cattle (*Bos taurus*), horses (*Equus caballus*), sheep (*Ovis aries*), goats (*Capra hircus*), swine (*Sus scrofa domesticus*), chickens (*Gallus gallus domesticus*), turkeys (*Meleagris gallopavo*), and ducks (*Anas*

Table 3–3. Use of fertilizer N, P, and K by farm production region in 1977 (USDA, 1981a).

Region	Fertilizer		
	N	P	K
		Gg	
Northeast	312.1	122.9	248.8
Lake	854.9	247.0	701.9
Corn Belt	2808.2	730.8	1795.6
Northern Plains	1551.9	230.4	119.1
Appalachian	635.6	208.5	481.6
Southeast	818.9	192.4	594.1
Delta	442.5	94.0	201.4
Southern Plains	911.9	154.9	100.1
Mountain	467.1	107.7	28.5
Pacific	815.0	130.3	94.1
Total	9618.1	2218.9	4365.2

platyrhynchos domesticus). About 90% of the approximately 158 000 Gg of manure generated annually, under both confined and unconfined conditions, is reportedly used as a production resource on land (USDA, 1978). About 96 000 Gg is excreted on pasture, rangeland, and cropland and thus is automatically returned to the land. About 73% of the 62 000 Gg produced under confined conditions is applied to the land. Although animal manure has long been used to improve soil tilth and fertility, it is important to recognize that there may be a number of constraints on the use of manure on land (Gilbertson et al., 1979). These can include the costs of collection, processing, transportation, and application; environmental considerations; and site characteristics. The site constraints, especially in terms of manure disposal, are thoroughly reviewed by Norstadt et al. (1977) and Witty and Flach (1977).

Data presented in Table 3–4 show the total annual production of manure by livestock by crop production region. The amounts of N, P, K, and C returned to the soil are shown from manure that is produced under confined conditions and which can be returned to cropland as a resource that can be managed. Manure was assumed to be 32% C on a dry weight basis (McCalla et al., 1977). In addition, an unquantified amount of the manure produced under unconfined conditions is also excreted on cropland and thus returned automatically. Therefore, our assumption that all of the manure produced under confinement was returned to cropland is an appropriate, but likely conservative estimate of nutrients and C that are returned to cropland from both confined and unconfined manure (Table 3–4). The author's estimates do not differ greatly from Power and Papendick's (1985) estimates made previously.

Crop Residues—Crop residues include stems, leaves, roots, chaff, and other plant parts that remain after agricultural crops are harvested or grazed. According to a recent USDA survey (USDA, 1978), about 390 000 Gg of crop residues are produced in the continental USA annually by 15

Table 3–4. Total annual production of animal manure and estimated return of N, P, K, and C to the soil from manure produced under confined conditions by farm production region (USDA, 1978).

Region	Total annual production	Returned to the soil			
		N	P	K	C
	Gg	Gg			
Northeast	9 644	137.4	83.2	141.4	1 986.5
Lake	14 334	145.0	101.5	238.7	3 285.4
Corn Belt	28 253	127.8	94.6	183.8	2 197.5
Northern Plains	20 488	49.0	50.8	118.5	1 438.0
Appalachian	13 808	58.8	46.5	73.2	955.0
Southeast	11 188	49.8	47.5	51.1	671.7
Delta	8 196	32.9	33.9	34.6	440.4
Southern Plains	24 061	61.0	45.7	99.8	1 241.6
Mountain	16 127	67.4	33.1	82.2	1 066.6
Pacific	12 034	79.8	52.0	104.3	1 462.0
Total	158 133	808.9	588.8	1 127.6	14 744.6

major cultivated crops; this residue accounts for about 80% of the total crop residues produced. Larson et al. (1978) estimated that about 280 000 Gg of crop residues were produced by 10 major crops. Three major crops—field corn, soybean, and wheat—produce about 70% of all crop residues (USDA, 1978). The quantities of residues and amounts of N, P, K, and C produced and returned to the soil are estimated by multiplying the total grain (or crop) production by a grain (or crop) to residue weight ratio (Table 3–5) (USDA, 1978). Crop residues were assumed to be 40% C on a dry weight basis (Parr and Papendick, 1978).

Disposition of crop residues can include feeding to animals, use as fuel, return to the soil, collection and selling, and wasted (e.g., burned in place) (Stanford Research Institute, 1976). On a national basis, about 70% of the residues and nutrients in them are returned to the soil, mostly at the production site. If livestock bedding wastes were included in Table 3–5, about 5% more crop residue nutrients would be accounted for on a national basis as being returned to the soil. The greatest quantities of bedding wastes result from dairy operations. Crop roots are also an important crop residue, but again are not included in Table 3–5.

Symbiotic Nitrogen Fixation—For this discussion, symbiotically fixed N returned to cropland includes that returned by the major seed legumes and by one forage legume (alfalfa, *Medicago sativa* L.). Table 3–5 shows that about 3 000 Gg of N are returned to the soil with crop residues each year. Of that amount, seed legumes (soybean, *Glycine max* L. Merr.; dry bean, *Phaseolus* spp., and peanut, *Arachis hypogaea* L.) provide about 24% nationally or 720 Gg; of that amount, soybean account for 96% of the legume-N (USDA, 1978).

A number of questions exist concerning the significance of N_2 fixation by soybean. In Illinois, Johnson et al. (1974) showed that the percent of total N in soybean plants derived from symbiotic fixation decreased from about 48% when no fertilizer-N was added down to about 10% with the

Table 3–5. Production of crop residues and return of N, P, K, and C to the soil by farm production region in 1977 (USDA, 1978).

Region	Total annual production	Returned to the soil			
		N	P	K	C
	Gg			Gg	
Northeast	10 145	76.7	10.6	85.4	2 694.6
Lake	40 365	294.0	42.4	332.8	10 462.6
Corn Belt	156 956	1 342.4	174.9	1 178.8	38 548.3
Northern Plains	73 960	484.9	63.2	640.8	22 395.2
Appalachian	15 676	161.2	19.3	132.4	4 376.6
Southeast	12 274	136.3	17.0	94.6	2 901.7
Delta	13 718	84.4	17.9	101.9	3 451.2
Southern Plains	29 691	181.5	22.4	219.5	7 589.0
Mountain	18 146	115.9	15.4	181.4	6 474.4
Pacific	18 337	109.2	13.3	163.6	5 977.9
Total	389 268	2 986.5	396.4	3 131.3	104 871.5

addition of 224 to 448 kg ha^{-1} of fertilizer N. Weber (1966) reported that with good growth conditions about 40% (72 kg ha^{-1}) of the N for soybean was symbiotically fixed on Midwestern soils. Ham et al. (1975) determined for Minnesota soils that soybean N$_2$ fixation provided only 34% and later Ham and Caldwell (1978) determined that symbiotic fixation by soybean provided only 25 to 27% of their N. More recently, Thurlow and Hiltbold (1985) indicate that estimates for Midwestern soils are too low for conditions in Alabama where 70% or more of the N for soybean is derived from the atmosphere. Hunt et al. (1985) for a 2-yr period in South Carolina estimated that the percentage of N supplied by N$_2$ fixation under either conventional or conservation tillage ranged between 49 and 67% for non-irrigated soybean. Under irrigation, Matheny and Hunt (1983) in South Carolina and Bezdicek et al. (1978) in Washington have reported that N$_2$ fixation by soybean accounted for up to 91 and 83% of total plant N, respectively.

In addition to questions concerning the percentage of the N that soybean symbiotically fixes from the atmosphere, there is also a question of estimation of amounts of soybean residues using straw to grain ratio. A straw to grain ratio of about 1.5:1.0 for soybean at harvest has been generally accepted in the literature (Larson et al., 1978). However, where ground litter is collected as it drops and is added back to the amount of standing plant material as part of the straw component, this ratio can change considerably. Hunt et al. (unpublished data) recently measured the straw to grain ratios, with ground litter included, to equal 2.7, 4.3, and 2.5 for three cultivars, respectively. Corresponding values for straw to grain ratio without the ground litter component were 1.2, 2.3, and 1.4 for the same three cultivars, respectively.

Based upon the above discussion, estimates can be made of the amounts of N reported in Table 3–3 that are derived from N$_2$-fixation. Generally accepted values for symbiotically fixed N would appear to be about 40 and 70% of the total under Midwestern (Weber, 1966) and Southeastern (Thurlow and Hiltbold, 1985) conditions, respectively. Assuming that Midwestern conditions are most similar to the Corn Belt, Northern Plains, Southern Plains, and Pacific Northwest tillage management regions (Fig. 3–2), and that Southeastern conditions most similar to Coastal Plains, Piedmont, and Eastern uplands; then approximately 72 and 28% of the soybean grown in the USA (USDA, 1979) have about 40 and 70% of their N needs met by fixation, respectively. Based upon the above assumptions and by also assuming that fixation for dry bean and peanut plants are similar to soybean plants, N$_2$ fixation would amount to about 48% of the 720 Gg fixed nationally in legume residues reported in Table 3–5 (USDA, 1978). Thus, about 350 Gg of N yr^{-1} or about 12% of the amount of N returned by the crop residues reported in Table 3–5 are from biological N$_2$ fixation by seed legumes.

Except for alfalfa, a major reduction in the use of forage legumes for biological N$_2$-fixation in cropping systems of the USA has occurred since about 1955 (Power, 1981; Power and Papendick, 1985). However, the feasibility of using legumes in cropping systems is receiving renewed interest (Ebelhar et al., 1984; Jones et al., 1983; Martin and Touchton, 1983; Power et al., 1983). Estimates of 7200 Gg of total annual biological N_2 fixation for USA's agriculture have been made. The bulk of symbiotically fixed N is immobilized within the herbage of the legume plant itself. Thus, much of it is removed with the harvested crop. Recently, Heichel (1986) identified that in legume-nonlegume crop sequences, the amount of N returned to the soil for use by the nonleguminous crop depends upon (i) the quantity of legume residue returned to the soil, (ii) the content of symbiotically fixed N in the residues, and (iii) the availability of the legume residue N to the succeeding nonlegume. Thus, to gain N-additions from forage legumes for succeeding crops, the legume must be managed to return N to the soil. The N available for incorporation into the soil depends upon the time of the season when incorporation occurs (Heichel, 1986) and the proportion of N-rich herbage that is incorporated into the soil as compared to N-poor crown and roots. The importance of the N-rich herbage is readily apparent from data in Table 3–6 from Heichel and Barnes (1984) which illustrates the relative amounts of N_2 fixed by alfalfa and removed from the soil for three cuttings of alfalfa during the seeding year. Heichel (1986) calculated that a typical alfalfa stand in the upper midwestern USA might contain no more than 6 kg of N ha^{-1} in root nodules.

For succeeding crops to benefit from symbiotically fixed N from forage legumes, they would presumably be grown in some type of crop rotation. Currently, much of the area used for growing forage legumes in the USA will be pastureland and hayland. Approximately 11 million ha of alfalfa were grown for hay in 1977 (USDA, 1979). Assuming that alfalfa stands are turned under at approximate 3-yr intervals, then about 3.6 million ha of alfalfa are turned under each year. The next assumption is an average fertilizer replacement value of 120 kg of N ha^{-1}, when alfalfa is managed as green manure or hay during its 3rd yr, plowed under in the fall and planted to a nonlegume the subsequent year (Heichel, 1986). Also, an average of 80% of the N in the alfalfa is assumed to have been

Table 3–6. Nitrogen budget for seeding year alfalfa showing the allocation of symbiotically fixed- and soil-derived N among plant parts (Heichel and Barnes, 1984).

Nitrogen budget component	Seeding year harvests		
	First (12 July)	Second (30 Aug.)	Third (20 Oct.)
Herbage yield (kg ha^{-1})	3.503	3.054	1.156
Total N yield (herbage, crown and roots) (kg of N ha^{-1})	118	127	59
Total N_2 fixed (kg of N ha^{-1})	57	102	34
Herbage	52	74	22
Roots and crown	5	28	12
Nitrogen from soil (kg of N ha^{-1})	61	25	25
Herbage	54	18	16
Roots and crown	7	7	9

symbiotically fixed (Heichel et al., 1984). Then, based upon the above assumptions, about 350 Gg of symbiotically fixed N yr^{-1} are returned annually to cropland from alfalfa in the USA.

Although other legumes are locally important, the assumption can likely be made (Power and Papendick, 1985) that the majority of forage legume N returned to cropland will be from alfalfa with other forage legumes accounting for some additional amounts; that does not result in a major change in the interpretation made here. The use of forage legumes in cropping systems is smaller than the potential appears to be. Reasons may be that they detract both space and time for the production of row crops. Also, the current widespread use of fertilizer N has greatly diminished the need to utilize legumes in cropping systems. Irrespective, there are new major opportunities to utilize forage as well as grain legumes as cover crops or catch crops in rotation with grain crops. Improved conservation tillage practices including no-till planting of a grain crop directly into a legume cover crop provides soil erosion control during periods of high erosion hazard. Additionally, it has the potential for reducing fertilizer N needs for the grain crop. For example, Ebelhar et al. (1984) estimated that hairy vetch (*Vicia villosa* Roth) in Kentucky supplied symbiotically fixed N equivalent to 90 to 100 kg ha^{-1} fertilizer N annually to maize while also serving as a winter cover crop to help provide soil erosion control.

Nutrient Resource Summary

The data from Tables 3–3, 3–4, 3–5, and from the discussions of the amounts of symbiotically fixed N added to the soil from soybean and alfalfa are summarized in Fig. 3–6. The authors recognize that these data are not complete in that the residues from a number of other crops produced on a limited scale are not included. Also, nutrient recycling from crop roots and livestock bedding wastes are not considered. Other

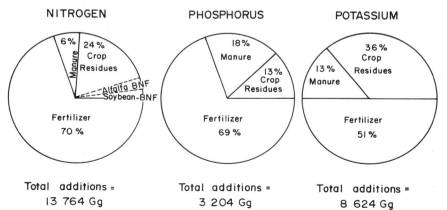

Fig. 3–6. Additions of N, P, and K to cropland soils of the USA from fertilizers and by return of crop residues and manure.

sources of nutrient additions such as from food processing wastes, municipal refuse, industrial organic wastes, sewage sludge and septage, and logging and wood manufacturing wastes are not included since they are generally considered of minor importance nationally (USDA, 1978). Irrespective, the relative amounts of nutrients added or returned to cropland soils from fertilizers, manures, and crop residues are shown. Such nutrients are important since a management choice is made whether to add these nutrients to the soil or not (fertilizers), or to return them to the soil, or remove them (i.e., crop residues) by burning, selling, or for some other purpose.

INFLUENCE OF CONSERVATION PRACTICES

Conservation practices influence the fate of plant nutrients in cropping systems. The degree of influence of conservation practices on the amounts of plant nutrients removed with the harvested crop, lost to the environment, or remaining in the soil for subsequent crop use (Fig. 3–3) needs to be evaluated. The following discussion is focused on situations where conservation practices are currently known to have a major impact and where sufficient data exist for an in-depth discussion. The evaluation will not be in-depth if conservation practices are only of minor importance or insufficient data exist.

Nutrient Losses

Soil Erosion

Soil erosion is perhaps the primary conservation problem on about one-half of U.S. cultivated croplands (Larson, 1981). A majority of the organic matter and available plant nutrients are near the soil surface and therefore are highly vulnerable to soil erosion. Off-site effects (Clark et al., 1985) as well as on-site effects of soil erosion are major national concerns. Loss of plant nutrients is a major consequence of soil erosion. Recent estimates of cropland losses of N, P, and K nationally are shown in Table 3–7 (Larson et al., 1983).

These data allow comparisons between amounts of plant nutrients from the various sources discussed previously and the amounts in eroded sediments. Data from Fig. 3–6 and Table 3–7 show that the ratio between total additions of N, P, and K and the total amounts of N, P, and K in eroded sediments are about 1.4:1.0, 1.9:1.0, and 0.2:1.0; respectively. Such a comparison does not consider removal of nutrients in harvested crops, leaching losses, surface runoff losses, or gaseous losses. Neither does it consider where eroded sediments and their associated nutrients are deposited, nor inherent fertility of the soil. Irrespective, these ratios indicate that the USA, on a national basis, is adding or returning more N and P but less K to the soil than is associated with eroded sediments. Over the long term, the relative differences between addition and return vs. erosion

Table 3–7. Total and available N, P, and K in eroded sediments (Larson et al., 1983).

Region	Nitrogen		Phosphorus		Potassium	
	Total	Available	Total	Available	Total	Available
			Gg			
Northeastern	300	55	75	1.5	2 252	45
Lake	622	114	107	2.1	3 643	73
Corn Belt	4 360	802	624	12.5	24 959	499
Northern Plains	2 068	380	293	5.9	11 711	234
Appalachian	676	124	169	3.4	3 381	67
Southeastern	202	37	101	2.0	1 007	20
Delta	478	88	141	2.8	4 220	84
Southern Plains	512	94	101	2.0	3 043	61
Mountain	176	32	64	1.3	2 550	51
Pacific	100	18	29	0.6	1 154	23
Total	9 494	1 744	1 704	34.1	57 920	1 158

of K appears to be of concern. However, the major crop production regions where the amounts of K (and N) are greatest in eroded sediments are the Corn Belt and the Northern Plains. In both regions, the soils have medium to high base supply and are relatively rich in organic matter. The ratios of total additions to total amounts in eroded sediments of N, P, and K for the Corn Belt and Northern Plains are about 1.0:1.0, 1.6:1.0, and 0.1:1.0 and 1.0:1.0, 1.2:1.0, and 0.1:1.0; respectively. Thus, ratios for both regions are lower than the national averages and it might be assumed that an overall net loss of N, P, and K may be occurring when the removal in the harvested crop, soil erosion, and other nutrient losses are considered collectively. Comparisons made at a regional or local level allow improved consideration of soil properties and overall interpretations. Also, conservation efforts need to be concentrated where erosion damage is greatest and not necessarily where the greatest amount of soil erosion occurs.

Soil Conservation and Organic Matter

Organic matter is primarily C (about 58% by weight) with a large reservoir of essential plant nutrients contained in it. Soil organic matter is also generally associated with the finer and more reactive clay and silt fractions of soils. Its proximity, and concentration near the soil surface (usually in the top 25 cm or less) and close association with plant nutrients in the soil, makes the erosion of soil organic matter a strong indicator of overall plant nutrient losses resulting from soil erosion. Thus, the effectiveness of soil conservation practices can also be evaluated based upon the amount of soil organic matter (organic carbon) associated with eroded sediments.

Tillage Effects—We used Land Resource Region M (USDA, 1981b) to demonstrate the effect of tillage on organic matter and nutrient losses due to water erosion. Land Resource Region M lies almost entirely within the North Central region and corresponds exactly to the Corn Belt Tillage

Management region identified by Allmaras et al. (1985) (Fig. 3–2 and Table 3–2). The following calculation procedures illustrate how soil survey and other data were used in conjunction with the Universal Soil Loss Equation (USLE), (Wischmeier and Smith, 1965, 1978) to calculate the effect of tillage on organic matter and nutrient losses due to water erosion.

Lindstrom et al. (1981) had previously assigned crop rotations to soil series and to slope-gradient classifications for determination of the cropping management (C) factor of the USLE. The assignment was based upon crop production statistics. The authors considered soil loss for three residue and tillage treatments. They are listed below.

1. Conventional tillage—No crop residue remaining on soil surface; equivalent to full moldboard plow, spring disk, and harrow.
2. Conservation tillage—3920 kg ha^{-1} of crop residue initially remaining on soil surface; subsurface tillage (chisel), 66% surface residue coverage.
3. No-till—3920 kg ha^{-1} of crop residue initially remaining on soil surface; 90% surface residue coverage.

The area of each soil association (Technical Committee on Soil Survey, 1960) by state and MLRA were determined and matched with the organic carbon content in the top 20 cm of soil as Franzmeier et al. (1985) reported for that soil association. A soil density of 1.4 g cc^{-1} was assumed for all soils to convert the organic carbon content Franzmeier et al. (1985) reported to percent organic carbon. Cultivated area, percentage of that area in each of four slope gradients (0–2, 3–5, 6–12, and >12%), and calculated soil loss rates by tillage treatment were obtained from Lindstrom et al. (1981). Next, a weighted average percent organic carbon was computed by state and MLRA from the soil associations found in each MLRA, fractions of the total area in each soil association, and the organic carbon content reported for each soil association (Franzmeier et al., 1985). Total erosion was calculated using previously calculated soil loss rates (Lindstrom et al., 1981) for each of the tillage treatments and area under each of the slope gradients. Organic carbon in eroded sediments was calculated by multiplying the weighted average percent organic carbon by soil erosion for each tillage treatment. Data from each MLRA (by state) were aggregated into totals for the Corn Belt Tillage Management Region. Organic nitrogen and organic phosphorus in eroded sediments were calculated assuming a ratio of organic carbon/organic nitrogen/organic phosphorous of 110:9:1 (Allison, 1973).

Since soil resource assessments and other activities associated with soil conservation are generally poorly correlated with soil testing activities, estimates that might be made of plant-available nutrients associated with eroded sediments are less reliable than are estimates of organic nutrients. Organic phosphorous, like organic nitrogen is concentrated in topsoil. Both are subject to mineralization and under favorable conditions supply a significant part of the N and P that plants needed. Therefore, the following discussion will be for organic nitrogen and phosphorous which will not always be closely correlated to plant-available N and P.

However, the loss of organic nutrients by erosion should be a good partial measure of overall nutrient losses.

Although loss of organic carbon (organic matter) is a direct function of soil loss, it is not a linear function. Since eroded materials frequently differ in composition from the original soil, the loss of nutrients may be expressed in terms of an enrichment ratio (ER) which is the ratio of the concentration of element in eroded soil material divided by the concentration of element in soil from which eroded soil material originated (Barrows and Kilmer, 1963). Organic matter is reported to have an average ER of 2.1 (Barrows and Kilmer, 1963). The value of ER reported for soil organic matter was assumed to be directly proportional for organic carbon with the constant organic carbon, nitrogen, and phosphorous ratio remaining at 110:9:1 (Allison, 1973). Separate enrichment ratios for N and P were not used since they are reported for total N and P rather than organic nitrogen and phosphorous (Barrows and Kilmer, 1963). Aggregation of the above calculations for Land Resource Region M are shown in Table 3–8.

Comparison of the amounts of organic carbon in eroded sediments (Table 3–8) for conventional tillage (49 000 Gg) to the total of that returned in animal manure (Table 3–4) and crop residue (Table 3–5) of about 40 000 Gg, shows the relatively high magnitudes and the overall importance of the return and management of organic residues to the soil organic matter budget in general. Although organic carbon is not necessarily misplaced from the landscape, these calculations show the potential for erosion to continually deplete organic carbon and emphasizes the importance of developing and adopting conservation practices to maintain, if not increase, soil C levels.

Slope and Tillage Effects—If targeting of soil conservation practices to the most serious soil erosion problems within MLRAs is to be accomplished, an understanding of the soil organic carbon, slope conditions, and amount of organic carbon in eroded sediments that may occur is needed. The basic unit for our computations was the soil series by slope gradient classification obtained from the SCS Conservation Needs Inventory (USDA, 1971). The area (converted to a percent of the total area)

Table 3–8. Annual soil erosion and amounts of organic carbon, nitrogen, and phosphorus in eroded sediments in the Corn Belt Tillage Management Region (Land Resource Region M) as influenced by tillage treatment.

Tillage treatment	Soil erosion	Organic carbon	Organic nitrogen	Organic phosphorus
		In eroded sediments†		
		Gg		
Conventional tillage	1 396 855	49 380	4 040	450
Conservation tillage	575 499	20 630	1 690	190
No-till	436 622	15 890	1 300	140

†An enrichment ratio of 2.1 was used for these calculations.

within an MLRA for dominating slope gradient and soil series was multiplied by the mid-range value of percent organic matter content from SOILS-5 data sheets for that soil series. Soil series were obtained by overlaying MLRA boundaries (USDA, 1981b) over Soil Association boundaries found in the North Central region (Technical Committee on Soil Survey, 1960). Weighted average soil organic carbon contents for each slope category were obtained for the four MLRAs of Land Resource Region M found in Minnesota and are shown in Table 3–9.

Multiplication of the values in Table 3–9 by the cultivated land area, cultivated soil loss rates, and an enrichment ratio of 2.1 (Barrows and Kilmer, 1963) for each of the tillage treatments and slopes reported in Lindstrom et al. (1981) give amounts of organic carbon in eroded sediments by MLRA, as shown in Fig. 3–7. By then applying an organic carbon to organic nitrogen to organic phosphorous ratio of 110:9:1 (Allison, 1973), estimates of erosion of all three nutrients were made. Figure 3–7 shows both the potential off-site and potential on-site impacts of soil erosion within each of the MLRAs. On-site losses refer to the rate (kg ha^{-1} per year) of nutrient losses due to soil erosion, whereas off-site losses mean total amount of nutrient losses (i.e., rate of nutrient loss as kg ha^{-1} times the area of the MLRA). The USLE (Wischmeier and Smith, 1978) is an entrainment model that does not account for sediment load loss or movement distance. Therefore, our calculations do not show that sediments are transported off of the landscape or where they are deposited. The length of each horizontal bar for a particular slope by tillage practice in Fig. 3–7 can be compared to the three scales shown on the horizontal axis of the graph to determine potential off-site losses of organic carbon, nitrogen, and phosphorous associated with eroded sediments. To evaluate potential on-site damage, the numbers from top to bottom on the left inside part of each graph give the kg ha^{-1} per year of organic carbon, nitrogen, and phosphorous in eroded sediment as a function of slope and tillage practice.

The high degree of load loss that might be expected, especially for the 0 to 2 and 3 to 5% slope conditions, is important to recognize in the following discussion of Fig. 3–7. In terms of nutrients associated with eroded sediments, the 0 to 2 and 3 to 5% slopes were generally highest (MN 102 and 103) because of the large areas of cultivated land and the

Table 3–9. Calculated average percent organic carbon in topsoil by slope category in Minnesota MLRAs 102, 103, 104, and 105.†

Slope percent	Mn 102	Mn 103	Mn 104	Mn 105
	% organic carbon			
0–2	3.59	4.43	2.80	2.23
3–5	2.82	2.48	2.00	1.35
6–12	2.33	1.80	2.00	0.89
>12	1.74	0.65	0.64	0.62

†Calculated values in this table were evaluated for their appropriateness by Dr. R. H. Rust of the Minnesota Soil Survey Staff, St. Paul (1986 personal communication, St. Paul).

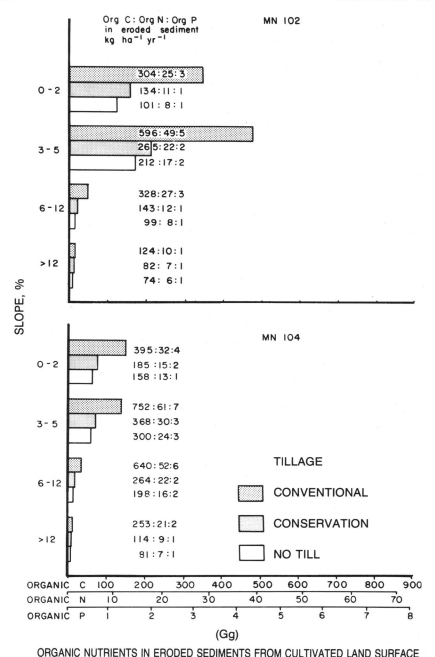

ORGANIC NUTRIENTS IN ERODED SEDIMENTS FROM CULTIVATED LAND SURFACE

Fig. 3–7. Organic nutrients in eroded sediments from cultivated land surface as a function of slope and tillage.

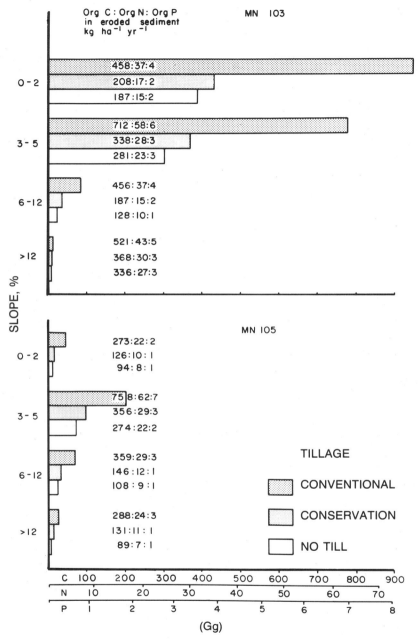

ORGANIC NUTRIENTS IN ERODED SEDIMENTS FROM CULTIVATED LAND SURFACE

Fig. 3–7. Continued.

higher soil organic matter contents. Organic carbon, nitrogen, and phosphorous associated with eroded sediments are generally much less for MN 104 and 105 than for MN 102 and 103 because of smaller cultivated area and/or lower soil organic matter contents. In addition, MN 105 and 104 have likely been cultivated for many more years than MN 103 and 102. The 3 to 5% slopes had the highest rates of organic carbon, nitrogen, and phosphorous loss per hectare (Fig. 3–7), while slopes of $> 12\%$ had the least. The lower rates of nutrient loss per hectare from the steepest slopes resulted from low organic matter content of soil from these slope conditions (Table 3–9).

Conservation tillage decreases the amount of organic nutrients associated with eroded sediments by about half; some additional decrease is obtained from no-till (Fig. 3–7). Where these practices are not sufficient to protect the soil resources or prevent loss of soil productivity, targeting of additional conservation practices may be necessary.

Other Losses

As shown in Fig. 3–3, other types of nutrient losses that conservation practices might affect are gaseous-, leaching-, and surface-runoff losses. Conservation tillage practices are of increasing interest because of their beneficial effects of reducing soil erosion and evaporative water losses. Reductions in the loss of soil organic matter and the retention of nutrients are additional benefits. No-till and other types of conservation tillage are reported to maintain larger reservoirs of potentially mineralizable nutrients, especially N, near the soil surface than does conventional tillage (Doran, 1980). Microbial populations with higher numbers of facultative anaerobes and denitrifiers have also been observed for no-till than for conventionally tilled soils (Doran, 1980). Thus, a greater potential for anaerobic metabolism and denitrification may occur with no-till than with conventional tillage.

Changes in soil water storage with conservation tillage are the result of four processes: suppressed overland flow, enhanced infiltration, more downward redistribution of water in the soil profile, and decreased evaporation. In a recent review, Allmaras et al. (1985) identified studies indicating that surface residues significantly reduce the velocity of overland flow so that infiltration is increased even if the infiltration rate is low.

To the degree that reduced tillage systems maintain larger reservoirs of potentially mineralizable N near the soil surface, the resulting question is whether increased infiltration associated with conservation tillage is conducive to leaching of soluble nutrients through and below the root zone. However, recent research by Elliott (Fort Collins, CO, 1986, personal communication) shows less leaching of nitrate (NO_3^-) under no-till than for stubble mulch or bare fallow in western Nebraska. He concluded that this decreased leaching was probably a result of the maintenance of soil structure and preferential flow of water down macropores or through inter-aggregate pore space. Higher moisture content, but less NO_3^- leaching in the no-till treatment indicated that NO_3^- was likely by-passed by preferential water flow through macropores. Nitrates resident within the aggregates and the need for diffusion of solutes out of aggregates may have resulted in lower NO_3^- concentrations in the draining soil water.

When soluble nutrients are on or near the soil surface (e.g., NO_3^-), they are usually leached into the soil by infiltration during the first part of a storm. Thus, the more infiltration there is before runoff begins, the lower the NO_3^- content of the runoff water. However, if the infiltrating water moves laterally and returns to the surface (interflow), its dissolved nutrient load is added to the overland flow. Stewart et al.'s (1976) report provides an overview on the control of water pollution from cropland. Surface runoff losses of dissolved nutrients are generally a small percentage of the total load of sediment transported nutrients (Gebhardt et al., 1985). Leaving crop residues on the land surface decreases runoff, but may not change the nutrient concentration of the runoff. For example, the effectiveness of conservation tillage for reducing soluble P in surface waters may depend on whether fertilizer is incorporated into the soil or not. If broadcast fertilizer is incorporated (or banded) into soil, reductions in P load are common under conservation tillage (Gebhardt et al., 1985). However, leaching of P from crop residues under no-till may increase the concentration of P in surface runoff waters while the soluble N load may not be affected at all (Stewart et al., 1976).

Nutrients Removed

The role of conservation practices for influencing the removal of plant nutrients in the harvested product (Fig. 3–3) must, to a large degree, be judged by the effect of conservation practices on increasing or decreasing crop yields. This assumes that the concentration of nutrients in the harvested product (e.g., grain) is essentially constant. Recently, Allmaras et al. (1985) reviewed research on crop yield responses to conservation tillage systems by TMR (Fig. 3–2). They identified major deterrents to the effective use of conservation tillage for maintaining or increasing crop yields as compared to conventional tillage including: weed problems; the lack of effective and/or adapted management inputs; non-availability of adapted crops or cultivars; insects and/or disease problems; and delayed planting or poor stands resulting from low soil temperatures, some soil properties (e.g., wet or high clay soils), or large amounts of crop residues.

Irrespective of the difficulties encountered in using conservation tillage, increases in crop yield are generally possible in all TMRs with significant increases possible in the Northern Great Plains. Where such increases do occur, the use of conservation practice(s) is resulting in increased removal of plant nutrients in the harvested crop. In addition, increased amounts of nutrients are taken up into crop residues to be recycled if the residues are returned to the soil.

Remaining Nutrients

The role of conservation practices for influencing the amount of plant nutrients remaining in the soil (Fig. 3–3) for subsequent crop use was described earlier in Eq. [1]. Maintenance of a fertile soil is closely associated with this reservoir of plant nutrients. Almost without exception, the reservoir of inorganic plant nutrients and those temporarily immobilized in various organic fractions within the soil are the prime sources of mineral nutrients taken up by the plants during the growing season.

As has been discussed earlier, organic matter improves soil fertility and a number of other desirable soil properties. Some cropping systems result in a build-up of soil organic matter, whereas others result in decreases. It is difficult to devise any cropping system that will prevent some decrease in the organic matter content in soils after they are plowed out of grass and put into grain production for the first time. However, after this initial decrease, various conservation and fertilization practices can be employed to prevent further decreases or even increase soil organic matter content (Power and Legg, 1978). The maintenance of or even restoration of soil organic matter levels may be one type of long-term measure of the effectiveness of conservation practices in maintaining soil productivity.

FUTURE ISSUES

Assessment Technologies

Targeting of appropriate soil conservation practices to maintain or improve soil productivity while controlling losses of plant nutrients by soil erosion, surface runoff, leaching, and perhaps even gaseous losses will require assessments. Guidelines will need to consider changes in soil chemical and physical properties across the landscape as influenced by conservation practices. At present there is only limited information on the effects of (i) landscape characteristics (slope and slope length), (ii) soil management practices (tillage), and (iii) crop management practices (double cropping, cover crop, and residue cover) on nutrient losses. If nutrient and organic matter data bases are available, then at least data bases, such as the Natural Resources Inventory (USDA, 1982), could be used to assess the changes in chemical aspects of soil productivity due to erosion. However, methods still need to be devised to improve assessments of leaching, gaseous, and surface runoff losses. Such methods will help delineate landscape, soil, or crop management factors that if managed properly will help preserve inherent soil fertility and thus soil productivity. Also, such methods require the availability and use of data bases. Availability of data is inadequate and techniques difficult for broad-scale assessments of the affects of conservation practices on improving plant nutrient management for crop production. However, such evaluations are needed and can help to identify those conservation practices that are most effective. Improvement and transfer of conservation practices to other soils and other geographical regions requires certain types of additional data as follows.

Delineation of Tillage Management Regions

The TMRs are too general in terms of conservation needs and functions. Further assessment is needed of the subdividing TMRs based upon soil fertility constraints for adoption of conservation tillage practices.

Improved Soil Test Data Base

To assess the influence of conservation tillage practices on future plant nutrient management, the current fertility status of soils, updated as often as feasible, is needed. Inventory of inherent fertility status could be accomplished through national coordination of existing and future soil test data bases in terms of nutrient amounts and not percent of samples in various qualitative categories (i.e., high, medium, and low). Scientific shortcomings of the soil test data bases in reporting quantitative soil fertility is recognized. However, this type of data base is essential for assessing changes in fertility status due to changing soil and crop management practices such as the increasing use of conservation tillage.

Soil Organic Matter Data Base

The organic matter status of soil allows evaluation of changes in the status of this resource due to changes in management schemes. Except for the organic matter data by soil association for North Central states (Franzmeier et al., 1985), there is no unified data base on organic matter status of U.S. soils. Soil organic matter and inherent fertility status of soils is routinely assessed during soil survey undertakings. The authors feel a national effort is needed to synthesize organic matter and fertility status by soil type, soil association, and major land resource area.

Merging of Data Bases

Assessment of the changes in on-site soil fertility status and possible off-site damage potential as influenced by soil (slope and tillage) and crop (double-cropping, cover crop, and residue cover) factors will require the merging of a soil test data base with soil resource data bases, such as those containing soil organic matter status, soil survey, soil erosion data, and possibly others.

Control Technologies

Indigenous soil nutrient resources, as well as nutrients that are added (fertilizer) or returned (crop residues, manures, etc.) to the soil must all be effectively managed. It is encouraging to observe the trend of increased use of conservation tillage for cropland in the USA. However, the complex interactions of conservation practices with improved plant nutrient management are still poorly understood. Even through conservation tillage has been the primary conservation practice that we have discussed, the use of a number of other water erosion (Laflen et al., 1985) and wind erosion (Fryrear and Sidmore, 1985) control practices may also be necessary. Much continued research is needed on the joint goals of continuing to improve crop yields while minimizing environmental degradation.

Increased use of conservation practices, especially conservation tillage, is helping to alleviate the effects of soil erosion on losses of nutrients and organic carbon from cropland. However, the overall effects of conservation practices on other types of losses, such as from gaseous losses, leaching, and surface runoff of dissolved nutrients and C is much less well understood. Even though these other types of nutrient and organic carbon losses may be assumed to be small relative to those resulting from soil erosion, their on-site and off-site effects (e.g., N leaching into groundwater) may result in significant environmental degradation. Nonetheless, the development of improved conservation practices to decrease the total loss of on-site nutrients is a necessary and worthwhile goal and requires a full understanding of the role of soil fertility and organic matter as critical components of production systems.

REFERENCES

Adams, J.R., L.B. Nelson, and D.B. Ibach. 1958. Crop-use patterns of fertilizer in the United States. Reprinted from Croplife. 18 August–13 October.

Allison, F.E. 1973. Soil organic matter and its role in crop production. Elsevier Scientific Publishing Co., New York.

Allmaras, R.R., P.W. Unger, and D.W. Wilkins. 1985. Conservation tillage systems and soil productivity. p. 357–412. In R.F. Follett and B.A. Stewart (ed.) Soil erosion and crop productivity. American Society of Agronomy, Crop Science Society of America, Soil Science Society of America, Madison, WI.

Barrows, H.L., and V.J. Kilmer. 1963. Plant nutrient losses from soils by water erosion. Adv. Agron. 15:303–316.

Bezdicek, D.F., D.W. Evans, B. Abede, and R.E. Witters. 1978. Evaluation of peat and granular inoculum for soybeans' yield and N-fixation under irrigation. Agron. J. 70:865–868.

Blevins, R.L., M.S. Smith, G.W. Thomas, and W.W. Frye. 1983. Influence of conservation tillage on soil properties. J. Soil Water Conserv. 36:301–304.

————, G.W. Thomas, and P.L. Cornelius. 1977. Influence of no-tillage and nitrogen fertilization on certain soil properties after 5 years of continuous corn. Agron.J. 69:383–386.

Broadbent, F.E.1985. Minimizing nitrogen losses in western valleys. p. 153–171. In Proceedings of plant nutrient use and the environment symposium. Kansas City, MO. 21–23 October. The Fertilizer Institute, Washington, DC.

Christensen, L.A., and R.S. Magleby. 1983. Conservation tillage use. J. Soil Water Conserv. 38:156–157.

Clark, E.H., II, J.A. Haverkamp, and W. Chapman. 1985. Eroding soils: the off-farm impacts. The Conservation Foundation, Washington, DC.

Conservation Tillage Information Center. 1985. National survey: conservation tillage practices (1984). National Association of Conservation Districts, Fort Wayne, IN.

Doran, J.W. 1980. Soil microbial and bio-chemical changes associated with reduced tillage. Soil Sci. Soc. Am. J. 44:765–771.

Ebelhar, S.A., W.W. Frye, and R.L. Blevins. 1984. Nitrogen from legume cover crops for no-tillage corn. Agron. J. 76:51–55.

Follett, R.F., and A. Bauer. 1986. Conservation-production-research issues. p. 93–100. *In* J. Maetzold and K. Alt (ed.) Forum on erosion productivity impact calculators. Soil Conservation Service Assessment and Planning Staff Rep. U.S. Department of Agriculture, Washington, DC.

————, and S.R. Wilkinson. 1985. Soil fertility and fertilization of forages. p. 304–317. *In* M.E. Heath et al. (ed.) Forages: the science of grassland agriculture. 4th ed. Iowa State University Press, Ames.

Franzmeier, D.P., G.D. Lemme, and R.J. Miles. 1985. Organic carbon in soils of North Central United States. Soil Sci. Soc. Am. J. 49:702–708.

Fryrear, D.W., and E.L. Skidmore. 1985. Methods for controlling wind erosion. p. 443–457. *In* R.F. Follett and B.A. Stewart (ed.) Soil erosion and crop productivity. American Society of Agronomy, Crop Science Society of America, Soil Science Society of America, Madison, WI.

Gebhardt, M.R., T.C. Daniel, E.E. Schweizer, and R.R. Allmaras. 1985. Conservation tillage. Science 230:625–630.

Gilbertson, C.B., F.A. Norstadt, A.C. Mathers, R.F. Holt, A.P. Barnett, T.M. McCalla, C.A. Onstad, R.A. Young, L.R. Shapler, L.A. Christensen, and D.L. Van Dyne. 1979. Animal waste utilization on cropland and pastureland: A manual for evaluating agronomic and environmental effects. USDA Utilization Rep. 6.

Ham, G.E., and A.C. Caldwell. 1978. Fertilizer placement effects on soybean seed yield, N_2 fixation, ^{33}P uptake. Agron. J. 70:779–783.

————, I.E. Liener, S.D. Evans, R.D. Frazier, and W.W. Nelson. 1975. Yield and composition of soybean seed as affected by N and S fertilization. Agron. J. 67:293–297.

Hargett, N.L., and J.T. Berry. 1985. 1984 fertilizer summary data. Bull. Y-180. National Fertilization Development Center, TVA, Muscle Shoals, AL.

Heichel, G.H. 1986. Legume nitrogen: symbiotic fixation and recovery by subsequent crops. *In* Z. Helsel (ed.) Fertilizers and pesticides. The Energy in World Agriculture Handb. Series. Elsevier Science Publishers, Amsterdam. (In press.)

————, and D.K. Barnes. 1984. Opportunities for meeting crop nitrogen needs from symbiotic nitrogen fixation. p. 49–59. *In* D. Bezdicek and J. Power (ed.) Organic farming: Current technology and its role in a sustainable agriculture. Spec. Pub. 46. American Society of Agronomy, Madison, WI.

————, ————, C.P. Vance, and K.I. Henjum. 1984. Di-nitrogen fixation, and N and dry matter partitioning during a 4-year alfalfa stand. Crop Sci. 24:811–815.

Hunt, P.G., T.A. Matheny, and A.G. Wollum. 1985. *Rhizobium japonicum* nodular occupancy, nitrogen accumulation, and yield for determinant soybean under conservation and conventional tillage. Agron. J. 77:579–584.

Johnson, J.W., L.F. Welch, and L.T. Kurtz. 1974. Soybeans' role in nitrogen balance. Ill. Res. 16:6–7.

Jones, J.N., Jr., H.D. Perry, E.L. Mathias, M.C. Carter, R.J. Wright, J.L. Hern, and O.L. Bennett. 1983. Conservation tillage in Appalachia. J. Soil Water Conserv. 38:219–221.

Laflen, J.M., R.E. Highfill, M. Amemiya, and C.K. Mutchler. 1985. Structures and methods for controlling water erosion. p. 31–442. *In* R.F. Follett and B.A. Stewart (ed.) Soil erosion and crop productivity. American Society of Agronomy, Crop Science Society of America, Soil Science Society of America, Madison, WI.

Lamb, J.A., G.A. Peterson, and C.R. Fenster. 1985. Wheat fallow tillage systems' effect on a newly cultivated grassland soils' nitrogen budget. Soil Sci. Soc. Am. J. 49:352–356.

Larson, W.E. 1981. Protecting the soil resource base. J. Soil and Water Conserv. 36:13–16.

————, C.E. Clapp, W.H. Pierre, and Y.B. Morochan. 1972. Effects of increasing amounts of organic residues or continuous corn II. Organic carbon, nitrogen, phosphorus, and sulfur. Agron. J. 64:204–208.

————, R.F. Holt, and C.W. Carlson. 1978. Residues for soil conservation. p. 1–15. *In* W.R. Oschwald (ed.) Crop residue management systems. Spec. Pub. 31. American Society of Agronomy, Madison, WI.

Larson, W.E., F.J. Pierce, and R.H. Dowdy. 1983. The threat of soil erosion to long-term production. Science 219:458–465.

Lessiter, F. 1973–1974 no-till farmer acreage survey. No-Till Farmer. March 1974.

————. 1974-1975 no-till farmer acreage survey. No-Till Farmer. March 1975.

————. 1976-1977 no-till farmer acreage survey. No-Till Farmer. March 1977.

————. 1978-1979 no-till farmer acreage survey. No-Till Farmer. March 1979.

————. 1980-1981 no-till farmer acreage survey. No-Till Farmer. March 1981.

————. 1982-1983 no-till farmer acreage survey. No-Till Farmer. March 1983.

————. 1984-1985 no-till farmer acreage survey. No-Till Farmer. March 1985.

Lindstrom, M.J., S.C. Gupta, C.A. Onstad, R.F. Holt, and W.E. Larson. 1981. Crop residue removal and tillage—effects on soil erosion and nutrient loss in the Corn Belt. USDA Agric. Information Bull. 442.

Lucas, R.E., J.B. Holtman, and L.G. Connor. 1977. Soil carbon dynamics and cropping practices. p. 333-351. In W. Lockeretz (ed.) Agriculture and energy. Academic Press, New York.

Magleby, R., D. Gadsby, D. Colacicco, and J. Thigpen. 1985. Trends in conservation tillage use. J. Soil Water Conserv. 40:274-276.

Mannering, J.V., and C.R. Fenster. 1983. What is conservation tillage? J. Soil Water Conserv. 38:141–143.

Martin, G.W., and J.T. Touchton. 1983. Legumes in cover crop and source of nitrogen. J. Soil Water Conserv. 38:214-216.

Matheny, T.A., and P.G. Hunt. 1983. Effects of irrigation on accumulation of soil and symbiotically fixed N by soybeans grown on Norfolk loamy sand. Agron. J. 75:719-722.

McCalla,T.M., J.R. Peterson, and C. Lue-Hing. 1977. Properties of agricultural and municipal wastes. p. 10–43. In L.F. Elliott and F.J. Stevenson (ed.) Soils for management of organic wastes and waste waters. Soil Science Society of America, American Society of Agronomy, Crop Science Society of America, Madison, WI.

Meisinger, J.J., V.A. Bendel, G. Stanford, and J.O. Legg. 1985. Nitrogen utilization of corn under minimal tillage. I. Four year results using labeled N fertilizer on Atlantic Coastal Plain soil. Agron. J. 77:602–611.

Norstadt, F.A., N.P. Swanson, and B.R. Sabey. 1977. Site design and management for utilization and disposal of organic wastes. p. 347–376. In L.F. Elliott and F.J. Stevenson (ed.) Soils for management of organic wastes and waste waters. Soil Science Society of America, American Society of Agronomy, Crop Science Society of America, Madison, WI.

Parr, J.F., and R.I. Papendick. 1978. Factors affecting the decomposition of crop residues by microorganisms. p. 101–129. In W.R. Oschwald (ed.) Crop residue management systems. American Society of Agronomy, Crop Science Society of America, Soil Science Society of America, Madison, WI.

Power, J.F. 1981. Nitrogen in cultivated ecosystems. p. 529–546. In F.E. Clark and T. Rosswall (ed.) Terrestrial nitrogen cycle—processes, ecosystem strategies and management impacts. Ecol. Bull. 33. Swedish Natural Science Resource Council, Stockholm, Sweden.

————, R.F. Follett, and G.E. Carlson. 1983. Legumes in conservation tillage systems: A research perspective. J. Soil Water Conserv. 38:217–218.

————, and J.O. Legg. 1978. Effect of crop residues on the soil chemical environment and nutrient availability. p. 85–100. In W.R. Oschwald (ed.) Crop residue management systems. American Society of Agronomy, Crop Science Society of America, Soil Science Society of America, Madison, WI.

————, and R.I. Papendick. 1985. Organic sources of nutrients. p. 503–520. In O.P. Engelstad (ed.) Fertilizer technology and use. 3rd ed. Soil Science Society of America, Madison, WI.

Rasmussen, P.E., R.R. Allmaras, C.R. Rohde, and N.C. Roager, Jr. 1980. Crop residue influences on soil carbon and nitrogen in wheat fallow system. Soil Sci. Soc. Am. J. 44:596–600.

Soil Conservation Society of America. 1982. Resource conservation glossary. SCSA, Ankeny, IA.

Stanford, G.O., O.L. Bennett, and J.F. Power. 1973. Conservation tillage practices and nutrient availability. p. 54–62. In Proc. Conservation tillage Symp. Soil Conservation Service of America, Ankeny, IA.

Stanford Research Institute. 1976. An evaluation of the use of agricultural residues as an energy feedstock. Vol. I and II. Prepared under the NSF Grant AER 74-18615A03.

Stewart, B.A., D.A. Woolhiser, W.H. Wischmeier, J.H. Caro, and M.H. Frere. 1976. Control of water pollution from cropland. Vol. 2—An overview. Joint ARS-EPA rep. EPA-600/2-75-0266 or ARS-H-5-2. U.S. Government Printing Office, Washington, DC.

Technical Committee on Soil Survey. 1960. Soils of the North Central Region of the United States. North Central Regional Pub. 76. Univ. of Wisconsin Agric. Exp. Stn. Bull. 544.

Thurlow, D.L., and A.E. Hiltbold. 1985. Dinitrogen fixation by soybeans in Alabama. Agron. J. 77:432–436.

USDA. 1971. Basic statistics—national inventory of soil and water conservation needs. Agric. Stn. Bull. 461, U.S. Department of Agriculture, Washington, DC.

————. 1978. Improving soils with organic wastes. Report to Congress in response to Section 1461 of the Food and Agric. Act of 1977 (P.L. 95–113). U.S. Department of Agriculture, Washington, DC.

————. 1979. Agricultural statistics. Prepared by the Economics, Statistics, and Cooperatives Service of USDA. U.S. Government Printing Office, Washington, DC.

————. 1981a. Fertilizer outlook and situation. USDA-ERS. FS-12.

————. 1981b. Land resource regions and major land resource areas of the United States. USDA-SCS Agric. Handb. 296. U.S. Government Printing Office, Washington, DC.

————. 1982. Basic statistics: 1977 National Resources Inventory. USDA Stn. Bull. 686. Washington, DC.

————. 1984. Cropland use and supply: Outlook and situation report. CUS-1. Economic Research Service. Washington, DC.

————. 1985a. The current financial condition of farmers and farm lenders. ERS Agric. Information Bull. 490.

————. 1985b. Inputs, outlook and situation report. IOS-7. Economic Research Service, Washington, DC.

Walsh, L.M. 1985. Plant nutrients and food production. p. 1–23. *In* Proc. Plant Nutrient Use andthe Environment Symp., Kansas City, MO. 21–23 October. The Fertilizer Institute, Washington, DC.

Weber, C.R. 1966. Modulating and nonmodulating soybean isolines: II. Response to applied nitrogen and modified soil conditions. Agron. J. 58:46–49.

Wischmeier, W.H., and D.D. Smith. 1965. Predicting rainfall-erosion losses from cropland east of the Rocky Mountains. USDA, Agric. Handb. 282. Washington, DC.

———— and ————. 1978. Predicting rainfall-erosion losses—a guide to conservation planning. Agric. Handb. 537. U.S. Department of Agriculture, Washington, DC.

Witty, J.E., and K.W. Flach. 1977. Site selection as related to utilization and disposal of organic wastes. p. 325–345. *In* L.F. Elliott and F.J. Stevenson (ed.) Soils for management of organic wastes and waste management. Soil Science Society of America, American Society of Agronomy, and Crop Science Society of America, Madison, WI.

4   Organic Matter Management and Utilization of Soil and Fertilizer Nutrients[1]

J. W. Doran and M. S. Smith[2]

This chapter presents relationships between soil management systems and soil and fertilizer nutrient availability. Nitrogen will be discussed more in depth than other nutrients because of its importance to crop production. The authors also want to better illustrate certain relationships. Emphasis is placed on systems in which organic matter, either plant residues or soil organic matter, is directly managed. Most examples will come from comparisons of tillage management systems; where the extent of residue incorporation varies and where soil organic matter is disturbed to varying degrees. Historically, approaches to regulating crop nutrient availability through organic matter management have involved: manipulation of existing soil organic matter through tillage or soil drainage; crop residue placement on or in soil and burning to enhance management operations; augmented in situ production using green manure, cover, and sod crops; and amendment with exogenous organic matter sources such as animal wastes and composts. A common factor in all these management practices, however, is the regulation of soil organic matter in (i) building reserves to enhance physical stability of soil and future nutrient supplies or (ii) the mining of organic reserves through biological oxidation and associated increases in plant-available N.

The native fertility of most forest and grassland soils in North America which were cleared of vegetation and cultivated for production of cash grain crops has declined significantly as soil organic matter was mined by crop removal without subsequent addition of plant and animal manures. As soil organic matter levels declined to 40 to 60% of their original levels, soil productivity declined, erosion losses of surface soil increased, and net mineralization of soil organic nitrogen fell below that needed for sustained grain crop production. An example of this relationship, with increasing time of cultivation, is that of the relative amount

[1] Contribution from the USDA-ARS in cooperation with the Nebraska Agric. Res. Div., Lincoln, and from the Kentucky Agric. Exp. Stn.

[2] Soil scientist, USDA-ARS, Univ. of Nebraska, Lincoln, NE 68583 and associate professor of soil microbiology, Univ. of Kentucky, Lexington, KY 40506

Copyright © 1987 Soil Science Society of America and American Society of Agronomy, 677 S. Segoe Rd., Madison, WI 53711, USA. *Soil Fertility and Organic Matter as Critical Components of Production Systems,* SSSA Spec. Pub. no. 19.

of plant-available N recycled from plant roots and residues increasing and that from the once fertile humus fraction declining (Fig. 4–1). To maintain yields with continuous cultivation supplemental N inputs from fertilizers, animal manures, or legumes are required.

Interpretation and prediction of the effects of various organic matter management systems on the availability of soil nutrients to crop plants depend on our understanding the unique roles played by the living and nonliving components of soil organic matter (Fig. 4–2). The living component, defined here to include soil microorganisms and fauna, make up a relatively small portion of total soil organic matter (1–8%). It functions, however, as an important catalyst for transformations of N and other nutrients. More recently the additional role of biomass as a "labile" source/sink for plant-available nutrients has been heavily emphasized, if not yet quantitatively defined. The majority of soil organic matter is contained in the nonliving component that the authors will define to include plant, animal, and microbial debris and soil humus. This component is important in defining the soil physical environment within which organisms function and in conferring stability to the soil ecosystem. This component has often been divided into labile vs. stable, active vs. inactive, or old vs. new fractions of differing chemical and biological availability. Although soil humus is clearly not uniform with regard to decomposition rates and the nature of its chemical and physical effects, it has not been demonstrated that discrete fractions can be identified in either a chemical or biological sense. Nevertheless, the designation of multiple, distinct pools appears to be a useful device for modeling and conceptualizing the roles and activity of soil organic matter. With this in mind, the authors suggest that the short- and long-term changes in nutrient cycling in response to organic management practices are associated with changes in the relative quantity and activity of labile and stable organic matter fractions. The activity and size of these pools will be dependent on climate, soil type, and management.

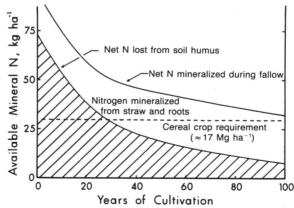

Fig. 4–1. Influence of cultivation time on relative mineralization from soil humus and wheat residue. After Campbell et al. (1976).

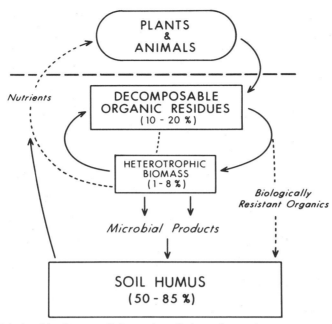

Fig. 4–2. Relationships between living and nonliving soil organic components, role in nutrient cycling, and relative proportion of total soil organic matter.

The value of defining and measuring soil organic fractions, as discussed above, in characterizing quality and activity of organic matter was demonstrated by Jannsen (1984). Using a model to separate the decomposition rates of young and old soil organic matter, he illustrated the effect of long-term amendment with mineral N, green manure, and animal manures on mineralization of soil N. As shown in Table 4–1 differences in N mineralization, after 25 yr of various management practices, were more closely related to the amounts of young organic matter than to the total N contents of the soil. Young soil organic matter being here defined as that accumulated by additions of crop residue and manure since management comparisons began and old organic matter being the remaining amount of total soil organic matter. This work indicates that the quality of soil organic matter can be influenced by different fertility management practices, and that qualitative changes may be more significant than gross quantitative changes.

The availability of supplemental and indigenous N to crop plants can also be greatly affected by management practices which involve soil tillage and crop residue management. The data presented in Fig. 4–3 illustrate the importance of degree of soil tillage on yield response of continuous maize (*Zea mays* L.) to fertilizer N. This also illustrates that constant, direct relationships between management practices and N availability should not be expected. These relationships are time, soil, and climate dependent. In Kentucky, during the first 9 yr cropping of a soil

Table 4–1. Influence of soil fertility management on soil density, organic matter (OM) pools, and associated N mineralization in surface soil (0–25 cm). After Janssen (1984).

Soil Property	Fertility management—Mineral fertilizers		
	Only	+ Green manure	+ Ley + Animal manure
	Mg/m³		
Bulk density	1.50	1.45	1.35
Organic matter	Mg/ha		
Total	78.4	81.9	86.9
Young	6.2	11.0	15.2
Old	72.2	70.9	71.7
	Mg N/ha		
Total soil N	4.0	4.4	4.6
Mineralization	mg N/g of total N		
6 weeks incubation	13	14	19
	kg N/ha per yr		
From young OM	34	74	108
From old OM	26	26	26

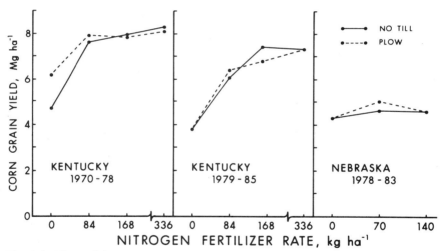

Fig. 4–3. Effects of tillage management and time on maize grain yield response to surface broadcast NH_4NO_3 at two locations. Kentucky data from Dr. R.L. Blevins, 1985, personal communication; Nebraska data after Wilhelm et. al. (1985).

that had previously been in bluegrass sod, maize yields with no tillage were significantly lower than those with plowing where no fertilizer N was applied but approximately equal when supplemental N was added. This has been attributed to greater immobilization of fertilizer N and less mineralization of soil N in no-till than in plowed soil (Kitur et al., 1984; Rice and Smith, 1984), although increased N loss may also be a factor. With time, however, yield differences between tillage practices disappeared. As the tillage systems are continued over many years, it may be assumed that each system approaches a new steady state in which total N reservoirs in the no-till soil exceed those in plowed soil (Rice et al., 1986). It is probable that the measured quantitative effects of tillage on soil N pools are accompanied by qualitative changes as well. The tillage effects on response to fertilizer N can also be influenced by climate, as indicated by the fact that at the Nebraska site there is no significant effect of tillage, tillage by N interactions, or duration of tillage management on maize yields. At this location, climatic factors have a greater effect on ultimate grain yield than either tillage management or N fertilizer application (Wilhelm et al., 1985). Another management difference between the two locations is a rye (*Secale cereale* L.) cover crop interseeded into maize at Kentucky while no cover crop was used in Nebraska.

To better manage the soil ecosystem, the authors would like to relate management practices or systems to the availability of plant nutrients (particularly N) in soil. Unfortunately, as already indicated, relationships between organic matter management and N availability are not always predictable, constant, or direct. To predict the influence of management practices on nutrient availability it is necessary to understand the mechanisms by which the processes controlling availability are regulated. In the case of organic matter management and N availability, three distinct (although sometimes interacting) categories of mechanisms can be described:

1. Direct physical/chemical effects where processes, such as NO_3 leaching and NH_3 volatilization, are altered by changes in soil structure, water relations, and/or soil surface chemistry.
2. Direct biological effects where processes, such as mineralization/ immobilization and N_2 fixation, are primarily altered via a change in energy status associated with changes in accessibility of plant and soil organic matter.
3. Indirect biological effects where processes such as denitrification/ nitrification, mineralization/immobilization, and plant root development and nutrient uptake are altered primarily via physical/ chemical changes associated with soil water status, aeration, and temperature.

The major point of such a categorization is to emphasize the importance of considering the dual role of soil organic matter both as a microbial substrate and as a determinant of the physical/chemical conditions in the soil environment. Thus, an important step in defining organic matter management effects on nutrient utilization and plant growth is understanding effects on the soil environment.

MANAGEMENT EFFECTS ON SOIL ENVIRONMENT/ BIOLOGICAL ACTIVITY

Agricultural production systems which manage organic matter directly and indirectly affect soil environmental factors which control the growth and activity of plants, microorganisms, and other living entities in soil (Fig. 4–4). Management which directly influences the placement and incorporation of plant and animal residues also controls the accessibility of such exogenous energy sources to soil microflora and fauna. Residue placement also indirectly influences soil environmental factors such as soil water and aeration status, soil temperature, and the relative predominance of certain organisms. These environmental factors, within the constraints set by climate and soils, play a major role in controlling and limiting biological processes determining the cycling of N and other elements. The degree of soil disturbance associated with management systems not only influences the aforementioned environmental factors but also the spacial habitat for biological activity through effects on soil bulk density, pore-size distribution, and aggregation. In trying to sort out the complex interactions between organic matter management and nutrient cycling, it is important to consider the synchrony of plants and microorganisms in responding to changes in the soil environment. This is a major consideration where organic matter reserves are augmented in situ through use of green manure and sod crops. A limitation of much research on effects of agricultural management practices on nutrient cycling and plant growth is the complexity of interactions between residue placement, soil disturbance, and type of crop. In the following section,

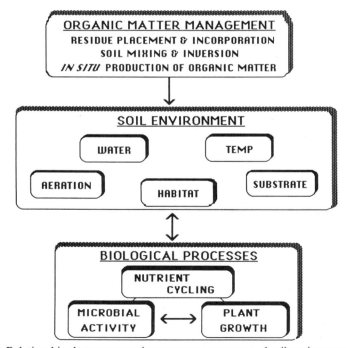

Fig. 4–4. Relationships between organic matter management and soil environmental factors which regulate biological processes and nutrient cycling.

the authors address the effects of management practices on C and N cycling and plant response by considering these areas separately.

Crop Residue Placement

Crop residues placed on the soil surface reflect light and insulate the soil thus reducing soil temperatures and evaporative losses of water (Bond and Willis, 1969; Gupta et al., 1983; Russell, 1939). The biological implications of residue related changes in soil water and temperature on plant development, nutrient uptake, and microbial activity are often climate and soil specific. In temperate regions during the spring warmup period, average soil temperatures at 100-mm depth can be 0.15 to 0.3 °C lower for each megagram of small grain residue applied to the soil surface (Allmaras et al., 1973). In a tropical environment, extreme temperatures often limit or reduce biological activity and application of 2 Mg/ha of crop residues can reduce surface soil (50-mm depth) temperature by as much as 8 °C (Lal, 1974). Thus, plant and microbial responses to temperature change resulting from surface residues depend greatly on climate, time of year, and relationships between temperature and biological response functions (Fig. 4–5). In a tropical climate, surface mulching may reduce soil temperature to a level more optimal for growth and activity of plants and microorganisms. When soils are warming in temperate climates, cooler temperatures with mulching often reduce biological activity.

In the subhumid environment of the western Corn Belt, grain yields of no-till maize and soybean increased 10% for each Mg/ha of crop residue (up to 5–6 Mg) maintained on the soil surface (Wilhelm et al., 1986). The major benefits of surface crop residues were through increased soil water storage and decreased temperatures during stressful midsummer periods. Also, the availability and N uptake from soil organic matter, crop residues, and applied fertilizer was higher with increasing surface crop residues presumably through creating a soil environment more favorable for microbial activity and N mineralization (Power et al., 1986).

Management related distribution of crop residues across the soil surface can also greatly influence soil biological activities and chemical properties. In eastern Nebraska, Doran (1980b) found that with stubble mulch management surface soil microbial populations between crop rows were 2 to 40 times greater than within crop rows. Between crop rows where residues were concentrated soil water contents, substrate supply, and soil pH were more optimal for microbial activity than within crop rows. Up to 145 kg of NO_3^--N/ha accumulated in surface soil between rows due in part to increased nitrification from more optimal water and pH regimes. This pool of N, however, was in dry soil and unavailable to the crop during much of the growing season and could be lost after harvest through leaching or denitrification. The potential for N conservation through microbial immobilization was implied but not studied.

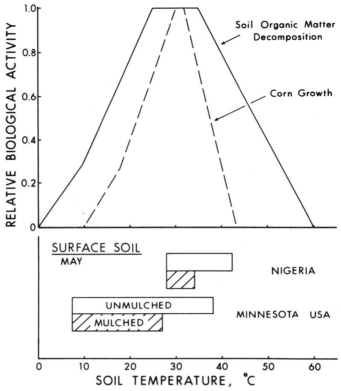

Fig. 4–5. Plant and microbial response to temperature and related influences of residue management and climate. Activity responses after Lehenbauer (1914) and Van Veen and Paul (1981); mulch effects after Gupta et al. (1983) and Lal (1976).

Crop residues, as well as influencing the soil environment for plant and microbial activity, also serve directly as substrate (C, N, etc.) which is converted to microbial biomass and soil organic matter. The rate of decomposition of residues is much slower when placed on the surface than when they are incorporated into soil (Brown and Dickey, 1970; Parker, 1962; Sain and Broadbent, 1977). Litter bag studies in Kentucky have shown that management of crop residues on the soil surface with no-tillage is a major factor in slower decomposition and accretion of surface organic matter as compared with moldboard plowing where residues are mixed into soil (Blevins et al., 1984). Climate also modifies the relative decomposition rates of surface and incorporated residues. In the semiarid Pacific Northwest, Douglas et al. (1980) measured decomposition losses over a 26-month period of 25 to 30 and 85% for wheat (*Triticum aestivum* L.) residues placed on or above the soil surface or buried at a 15-cm soil depth, respectively. In the warmer climate of the humid southeastern USA, rates of wheat residue decomposition (Fig. 4–6) were also greater with buried as compared to surface placement, but

WHEAT RESIDUE DECOMPOSITION

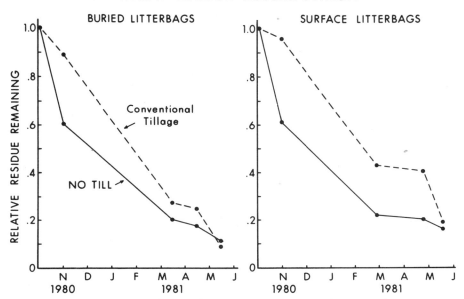

Fig. 4–6. Wheat residue loss from surface and buried litterbags with conventional and no-till management. After House et al. (1984).

in either case more than 80% had decomposed within an 8-month period (House et al., 1984).

The manner in which decomposition and the associated release or utilization of nutrients can be affected by residue placement is perhaps more interesting than differences in rates of decomposition. In Georgia, the stratification of crop residues, organic matter, and soil organisms with no-tillage management is a major mechanism for immobilization of N near the soil surface and slower recycling of N as compared with conventional tillage with the moldboard plow (House et al., 1984). Greater conservation of N in the organic form within the no-tillage ecosystem resulted from greater populations of weeds, soil insects, microarthropods, earthworms, and microorganisms which recycled nutrients within the soil/plant ecosystem. With conventional tillage biological diversity in surface soil was greatly reduced thus increasing the potential for leaching of nutrients released during decomposition of residues. In the Georgia study, rates of decomposition for wheat and soybean [*Glycine max* (L.) Merr.] residues were greater where buried; however, the overall seasonal decomposition was similar for no-tillage and conventional management. Decomposition of surface placed soybean residues with conventional tillage was significantly less than with no tillage.

Holland and Coleman (1986) identified changes in soil microclimate and microbial populations resulting from residue stratification as mechanisms by which organic matter and nutrients were conserved during residue decomposition with reduced tillage in the Western Great Plains. Soil fungal biomass was greater for surface compared with incorporated wheat straw. Holland and Coleman hypothesized that because fungi are believed to have a higher substrate utilization efficiency, a higher C/N ratio, a greater tolerance of variable moisture, and ability to extend to soil N reserves fungal-dominated decomposition of surface residues would result in greater C retention near the soil surface. Where crop residues were incorporated into soil a predominance of bacterial activity would result in greater C loss during decompositon and perhaps greater potential loss of available N.

Of the many biological processes influenced by tillage and associated residue management, mineralization/immobilization appears most important to nutrient cycling. Of particular interest is the potential for decreased mineralization of soil organic nitrogen and increased immobilization of surface fertilizer applications with reduced or no-tillage systems where residues and microorganisms are concentrated near the soil surface (Doran, 1980a; Cochran et al., 1980; Kitur et al., 1984). In England where crop residues are routinely burned before planting, Dowdell and Crees (1980) found no difference in N immobilization between no-tilled and plowed winter wheat. In Kentucky, however, where crop residues are retained and maize is planted directly into a chemically killed rye cover crop Rice and Smith (1984) found increased gross immobilization of nitrogen-15 (^{15}N) in surface no-tillage soils (Table 4–2). Although this observation is apparently explained, but only in part, by greater rates of organic/inorganic N turnover in no-tillage, they suggested immobilization is a major factor in reduced crop recovery of fertilizer N with no-tillage. Increased N immobilization was not observed in studies from drier climates in Nebraska and Saskatchewan (Carter and Rennie, 1984a, 1984b; Wilhelm et al., 1985). Again this illustrates the importance of understanding the modifying effects of environment on management responses.

Nitrogen fertilizer placement effects apparently are related to residue placement and tillage. Maize yields with no-tillage are often less at sub-

Table 4–2. Recovery of surface broadcast ^{15}N ammonium sulfate (168 kg of N/ha) in no-till and plowed soils 5 weeks after application. After Rice and Smith (1984).

Soil	Tillage	Fertilizer N recovered in soil N		
		Organic	Inorganic	Total
		kg N/ha		
Maury SiL				
	No-till	35**	79	114
	Plow	18	89	107
Cavode SiL				
	No-till	41*	31	72
	Plow	25	37	62
Tilsit SiL				
	No-till	22	32	54
	Plow	10	72*	82

*, **Significant difference between tillage treatment at P <0.05 and 0.01, respectively.

optimal rates of surface-applied N fertilization but at rates exceeding 100 kg of N/ha are the same or more than those with conventional tillage (Doran and Power, 1983; Meisinger et al., 1985; Thomas and Frye, 1984). Differences in N response have been attributed, in part, to increased potentials for volatilization of broadcast N and immobilization by greater microbial populations present in the surface of no-tillage soils. Placement of fertilizer N below the surface layer of no-tillage soils, where biomass levels and concentrations of high C/N organic materials are lower, can increase N availability, plant uptake, and ultimate crop yields (Table 4–3).

Soil Tillage—Organic Matter Management and N Cycling

Soil tillage is one of the oldest known management techniques to mine organic nutrient reserves for enhanced crop production. In cultivating virgin forest and grassland soils, decline of soil organic carbon and nitrogen, and corresponding increases in plant-available N, are greatest during the first 10 yr and decrease with time thereafter (Campbell et al., 1976; Giddens, 1957; Haas et al., 1957). Climate is an important modifier of tillage and plant-related changes in soil organic matter. In subhumid temperate climates, decline of soil organic nitrogen reserves in cultivated cropland is slowed by use of legumes or reduced tillage management practices (Bauer and Black, 1981; Tiessen et al., 1982). Higher biological activity in warmer, more humid climates reduces the effectiveness of green manure crops and resistant organic materials for increasing soil organic matter levels in cultivated soils (Giddens, 1957).

Table 4–3. Effect of N placement on grain yield of no-till maize. Sources: Maryland— V.A. Bandel, unpublished data, Georgia—Touchton and Hargrove (1982), Kentucky— W.W. Frye, (1982 personal communication); Illinois and Indiana—Mengel et al. (1982).

| Location, soil(s) | Nitrogen | | Average grain yields | | Increase for subsurface placement |
	Source	Rate	Surface broadcast	Subsurface applied	
		——— kg/ha ———	——— Mg/ha ———		%
Maryland					
Coastal SiL	UAN	134	6.2	7.8	26
Piedmont SiL	UAN	134	7.5	10.5	40
Coastal SiL	UAN	134	8.5	9.8	15
Coastal SiL	UAN	179	10.0	11.2	12
Georgia					
Piedmont SL	UAN	90	5.6	7.9	41
Piedmont SL	UAN	180	6.8	8.8	29
Kentucky					
Silt loam	Urea	80	6.7	8.0	19
	NH_4NO_3	80	7.6	9.8	29
Illinois					
Silt loam	UAN	165	4.8	6.4	33
Indiana					
3 silt loams	UAN	165	8.5	9.4	12

Recent studies of management systems and nutrient availability deal with mixed effects of soil disturbance and plant residue placement. In terms of understanding mechanisms, however, it seems appropriate to separate the two. Soil tillage affects major factors which limit or regulate biological activity—namely, water, O_2, temperature, the accessibility of C and N substrates, and the habitats for activity of plants, microflora, and microfauna. Tillage imposed changes in these limiting factors greatly influence cycling of C and N. In undisturbed plant/soil environments the soil organic matter content is fairly constant and much of the N released during organic matter oxidation is conserved in the cells of microorganisms, taken up by plants, or conserved as surface litter. Tillage alters this steady state through associated physical changes in the structure, aeration, soil water status, and availability of C and N as energy and nutrient sources of microorganisms and plants.

The nature and magnitude of soil physical responses to tillage often depend on previous cropping management and initial soil organic matter content. There is a strong inverse relationship between soil organic matter content and soil bulk density (Bauer, 1974; Russel, 1960) and cultivation of virgin soils usually results in increased bulk densities as organic matter levels decline with time (Cameron et al., 1981). In arable croplands, initially lower in soil organic matter, soil bulk densities with no-tillage or reduced tillage are usually greater than their plowed or cultivated counterparts (Griffith et al., 1977; Lal, 1976; Russell et al., 1975). Reduced tillage of previously cultivated soils not only reduces total pore space but also pore-size distributions. Non-tilled arable soils generally have more fine pores and fewer air-filled pores (pores > 30 μm) than plowed or cultivated soils (Ehlers, 1973; Russell et al., 1975; Van Ouwerkerk and Boone, 1970).

Degree of soil tillage can also profoundly influence O_2 dependent microbial transformations of C and N (Linn and Doran, 1984b). At six locations in the USA, Linn and Doran (1984a) found a greater relative predominance of anaerobic to aerobic organisms in the surface of no-tillage compared with plowed soils. This change was associated with wetter more compact conditions with reduced tillage. After addition of water, greater denitrification in no-tillage compared with plowed soils was highly correlated with less aerobic conditions. Other researchers have also found that, shortly after rainfall or irrigation, greater soil water content and reduced air-filled porosity are major factors for higher denitrification rates and gaseous losses of N in no-tillage as compared with plowed soils (Aulakh et al., 1982; Rice and Smith, 1982). The potential for less aerobic conditions with reduced tillage is site specific and depends on climate, soil, porosity and drainage characteristics, and quantity of crop residues maintained on the soil surface (Fig. 4–7). In the drier climate of western Nebraska, greater potential for denitrification with reduced tillage, as indicated by higher supplies of organic substrates and NO_3-N and greater numbers of denitrifying organisms, may rarely result in significant losses of N because soil aeration is only occasionally limited by excessive water

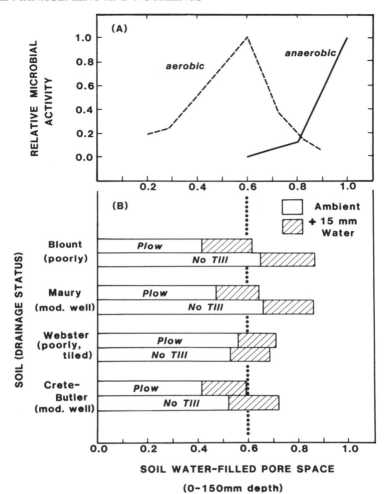

Fig. 4–7. Relationships between water-filled pore space, microbial activity, tillage management, and soil drainage status. After Mielke et al. (1986).

contents (Broder et al., 1984). Significant denitrification losses of N from reduced tillage soils may be limited to poorly or imperfectly drained soils (Blevins et al., 1984).

Aside from influencing the environment for biological activity, tillage also disrupts the physical arrangement of soil particles and facilitates microbial decomposition of organic substrates which are inaccessible in undisturbed soils (Rovira and Greacen, 1957; Van Veen and Paul, 1981). Cultivation, and associated changes in crop management and wetting and drying of soils, greatly reduces the number of soil macroaggregates (size >250 μm) which are stabilized by fungal hyphae and plant roots (Tisdall and Oades, 1982). The N in microbial and organic materials, which are normally occluded within macroaggregates and protected from biological oxidation in undisturbed soils, is mineralized after cultivation (Elliott, 1986). Organic matter characterization for clay soils in Australia which had lost up to 66% of their original C (and N) from cultivation revealed that the majority of organic matter in both cultivated and undisturbed native soils was present as alkyl forms which are normally considered susceptible to biological oxidation (Skjemstad et al., 1986). This again supports the importance of physical protection to organic matter accumulation in undisturbed soils.

Many studies of soil organic matter mineralization have compared management systems involving frequent, thorough disturbance of soil to systems with little or no soil disturbance. As indicated in Table 4–4 surface soil levels of organic matter, microbial biomass, and potentially mineralizable N are all significantly higher with no-tillage as compared with conventional tillage. As a result of increased residue and organic matter levels and more optimal water status, higher microbial biomass levels in surface no-tillage soils result in a greater reserve of N in potentially mineralizable forms. In reality, most farmers adopting modern minimum-tillage systems practice some periodic disturbance of soil. Tillage should be considered a powerful technique for managing stored soil nutrients. Although it will not be a simple task, because of the climate and soil specific effects of tillage, it should be possible to devise management systems which include rotational tillage to optimize the temporal or spatial correlation between crop needs and nutrient release or storage.

In Situ Production of Organic Carbon and Nutrients

Historically, legume cover crops, green crop and animal manures, and pasture rotations have been used to augment soil organic matter levels and supply N and other nutrients for crop production. Within the

Table 4–4. Effect of tillage management on soil water and organic matter contents and levels of microbial biomass and potentially mineralizable N. Sources: U.S.—Doran et al. (1985), Canada—Carter and Rennie (1982), England—Lynch and Panting (1980).

Location(s), depth in soil	Relative difference no-tillage vs. conventional tillage			
	Water content	Organic matter	Microbial biomass	Potential mineralizable N
——— cm ———	——————————————— % ———————————————			
USA (six sites)				
0–7.5	+22	+40	+58	+34
7.5–15	+3	−1	−2	−3
15–30	+5	−7	0	−7
Canada (four sites)				
0–5	−4	−11	+11	+59
5–10	+2	−2	+6	−25
10–20	−	−	−6	−
England (one site)				
0–5	+2	+16	+32	−

last 30 yr, however, many of these management approaches have been displaced by the increased use of crop monoculture, fertilizers, and pesticides. Concerns for reducing capital inputs, soil erosion losses, environmental degradation, and sustaining soil productivity have led to renewed interest in biological methods for supplying nutrients and building soil organic matter (Power and Papendick, 1985; USDA, 1980). This is typified by increased research on legume winter cover crops in parts of the country where climate permits. Although an old practice, much remains to be learned about how legumes fit into modern cropping systems and, at a less applied level, how these affect the soil ecosystem. The mechanism by which legumes input N to the plant/soil ecosystem is through alteration of the microbial energy status, i.e., by supplying substrate to symbiotic microbial partners which fix atmospheric N_2. Depending on climate, management, and the particular legume crop used a yearly fertilizer N equivalent of from 10 to 170 kg of N/ha is made available to nonlegumes in crop rotations and in organic farming systems provides a major part of the N required for crop production (Heichel and Barnes, 1984; Voss and Shrader, 1984).

Limited quantitative information is available on the fate of biologically fixed organic nitrogen and its transfer to succeeding grain crops in rotation with legumes. One would expect a tighter N budget in a cropping system relying on legumes and less denitrification and leaching losses of fixed organic nitrogen as compared to inorganic N fertilizer. However, the small amount of tracer data available suggests that recovery of crop N in following grain crops is rather low. Bartholomew (1971) summarized data on isotopic N recovery by crops following legumes. Recovery of N fixed in legumes ranged from 50 to 90% under controlled environmental conditions in the greenhouse but only 10 to 50% was recovered during the 1st yr by grain crops in the field. Only about 15% of the ^{15}N incorporated in medic clover residues in three Australian soils was taken up in the following wheat crop, and 75% of labeled N was found in soil organic matter (Ladd, 1981).

The low recovery of cover crop N may result from competition between plants and heterotrophic microorganisms for inorganic N and increased storage in soil organic nitrogen pools. The data presented in Table 4–5 suggests there is a potential for better management of the organic nitrogen in cover crops and a potential for using soil tillage to increase

Table 4–5. Influence of tillage and cover crop on maize grain yield and recovery of cover crop N in maize grain and stover. After Varco et al. (1985).

Tillage	Cover crop	Maize grain yield	Cover crop N recovered
		Mg/ha	%
No-tillage	Vetch	6.4	16
No-tillage	Rye	3.3	21
Plow	Vetch	6.9	30
Plow	Rye	5.0	31

N transfer to succeeding crops. Other management possibilities include timing of when the cover crop is killed, partial harvest of cover, and selection of crop species for optimum accumulation of C and N. Research indicates that enhanced nutrient (N) supply is not the only benefit of legumes as cover crops or in rotation with grain crops (Voss and Shrader, 1984). Their use in a cropping system may also alter soil structure, aeration, and water relations in soil and minimize soil erosion losses (Bezdicek, 1984; USDA, 1980). The long-term effects of using legume cover crops will also include protection of soil and environmental resources through root development and storage of nutrients between seasons when available nutrients are often lost from the soil ecosystem. The resultant effects of these changes on microbial nutrient transformations and sustained crop production needs further quantification.

CONCLUSION

Agricultural management systems control crop residue placement, soil disturbance, and the in situ production of soil organic matter. The resultant effects on soil physical properties, substrate availability, and organism habitat greatly influence plant, faunal, and microbial activities and the cycling of C, N, and other nutrients. Interactions between the soil physical, chemical, and biological characteristics with various systems of organic matter management are greatly influenced by climate, soil, and initial soil organic matter levels. Development of alternate management strategies for most efficient utilization of N and other elements requires better understanding of these interactions in the soil ecosystem. Several areas of future research and critical questions to answer emerge:

1. What are the long-term effects of management systems on organic nutrient reserves, nutrient availability, and microbial processes? We have presented some data (Fig. 4–3) which suggest that these relationships can change slowly following the imposition of new management techniques, yet most of the current information can be applied only to initial, perhaps transient conditions. For example, we do not know if no-tillage soils will continually immobilize more and mineralize less N than conventionally tilled soils.
2. How can we better define the function of soil organic matter? Both biological activity, mineralization rate for example, and physical/chemical function, as a determinant of soil structure for example, need to be considered. Can meaningful biological or chemical fractionation procedures be devised to better characterize organic matter qualitatively and quantitatively?
3. What is the optimum balance between organic matter exploitation as a source of nutrients and energy and conservation as a conditioner of the soil environment? Assuming that organic matter can be manipulated practically, how much and when should it be

changed? Of course, this is an economic and social question, as well as agronomic.

4. There is a great need for better methods of directly measuring N transformations in the field. Laboratory studies on disturbed soil samples are inadequate for assessing management effects; this is most apparent in tillage studies. Most current methods of measuring such processes as denitrification, soil N mineralization, or N_2 fixation in situ are either extremely time consuming, of controversial validity, or not widely applicable. Without reliable, convenient process measurements in the field we cannot hope to gain a quantitative level of understanding of management effects on nutrient availability.

5. A challenge to enhance our predictive modeling capabilities: i) to better quantitatively predict effects of specific organic matter management practices on the physical-chemical environment under varying climates and soil types and ii) to better quantitative understanding and modeling of the relationships between soil environment and rates/extent of nutrient transformations.

REFERENCES

Allmaras, R.R., A.L. Black, and R.W. Rickman. 1973. Tillage, soil environment and root growth. p. 62–86. *In* Conservation tillage, Proc. Nat. Conf., Des Moines, IA. 28-30 March. Soil Conservation Society of America, Ankeny, IA.

Aulakh, M.S., D.A. Rennie, and E.A. Paul. 1982. Gaseous nitrogen losses from cropped and summer-fallowed soils. Can. J. Soil Sci. 62:187–196.

Bartholomew, W.V. 1971. ^{15}N research on the availability and crop use of nitrogen. p. 1–20. *In* Nitrogen -15 in soil-plant studies. International Atomic Energy Agency, Vienna.

Bauer, A. 1974. Influence of soil organic matter on bulk density and available water capacity of soils. North Dakota Agric. Exp. Stn., Farm Res. 31:44–52.

----, and A.L. Black. 1981. Soil carbon, nitrogen, and bulk density comparisons in two cropland tillage systems after 25 years and in virgin grassland. Soil Sci. Soc. Am. J. 45:1166–1170.

Bezdicek, D.F. 1984. Organic farming: Current technology and its role in a sustainable agriculture. Am. Soc. Agron. Spec. Pub. 46. Madison, WI.

Blevins, R.L., M.S. Smith, and G.W. Thomas. 1984. Changes in soil properties under no-tillage. p. 190–230. *In* R.E. and S.H. Phillips (ed.) No-tillage agriculture. Van Nostrand Reinhold Co., New York.

Bond, J.J., and W.D. Willis. 1969. Soil water evaporation: surface residue rate and placement effects. Soil Sci. Soc. Am. Proc. 33:445–448.

Broder, M.W., J.W. Doran, G.A. Peterson, and C.R. Fenster. 1984. Fallow tillage influence on spring populations of soil nitrifiers, denitrifiers, and available nitrogen. Soil Sci. Soc. Am. J. 48:1060–1067.

Brown, P.L., and D.D. Dickey. 1970. Losses of wheat straw residue under simulated field conditions. Soil Sci. Am. Proc. 34:118–121.

Cameron, D.R., C. Shaykewich, E. de Jong, D. Chanasyk, M. Green, and D.W.L. Reed. 1981. Physical aspects of soil degradation. p. 186–255. *In* Proc., 18th Annual Alberta Soil Science Workshop, Edmonton, Alberta. 24-25 Feb.

Campbell, C.A., E.A. Paul, and W.B. McGill. 1976. Effect of cultivation and cropping on the amounts and forms of soil N. p. 9–101. *In* W.A. Rice (ed.) Proc. Western Can. Nitrogen Symp., Calgary, Alberta, Canada, 20-21 January. Alberta Agriculture, Edmonton, Alberta, Canada.

Carter, M.R., and D.A. Rennie. 1982. Changes in soil quality under zero tillage farming systems: distribution of microbial biomass and mineralizable C and N potentials. Can. J. Soil Sci. 62:587–597.

----, and ----. 1984a. Nitrogen transformations under zero and shallow tillage. Soil Sci. Soc. Am. J. 48:1077–1081.

----, and ----. 1984b. Crop utilization of placed and broadcast ^{15}N-urea fertilizer under zero and conventional tillage. Can. J. Soil Sci. 64:563–570.

Cochran, V.L., L.F. Elliott, and R.I. Papendick. 1980. Carbon and nitrogen movement from surface-applied wheat straw. Soil Sci. Soc. Am. J. 44:978–982.

Doran, J.W. 1980a. Microbial changes associated with residue management with reduced tillage. Soil Sci. Soc. Am. J. 44:518–524.

----. 1980b. Soil microbial and biochemical changes associated with reduced tillage. Soil Sci. Soc. Am. J. 44:765–771.

----, L.N. Mielke, and J.F. Power. 1985. Tillage-imposed changes in the agricultural ecosystem. p. 23. In Proc., 10th Conf., Reduced Tillage-Rational Use in Sustained Production. Int. Soil Tillage Res. Org., Guelph, Ontario, Canada. 8–12 July. Elsevier Science Publishing, Amsterdam.

----, and J.F. Power. 1983. The effects of tillage on the nitrogen cycle in corn and wheat production. p. 441–455. In R. Lowrance et al. (ed.) Nutrient cycling in agricultural ecosystems. Univ. Georgia Coll. Agric. Spec. Pub. 23. Athens, GA.

Douglas, C.L., Jr., R.R. Allmaras, P.E. Rasmussen, R.E. Ramig, and N.C. Roager, Jr. 1980. Wheat straw composition and placement effects on decomposition in dryland agriculture of the Pacific Northwest. Soil Sci. Soc. Am. J. 44:833–837.

Dowdell, R.J., and R. Crees. 1980. The uptake of ^{15}N-labeled fertilizer by winter wheat and its immobilization in a clay soil after direct drilling or plowing. J. Sci. Food Agric. 31:992–996.

Ehlers, W. 1973. Total porosity and pore size distribution in untilled and tilled soils. Z. Pflanzenernaehr. Bodenkd. 134:193–207.

Elliott, E.T. 1986. Aggregate structure and carbon, nitrogen, and phosphorus in native and cultivated soils. Soil Sci. Soc. Am. J. 50: 627–633.

Giddens, J. 1957. Rate of loss of carbon from Georgia soils. Soil Sci. Soc. Am. Proc. 21:513–515.

Griffith, D.R., J.V. Mannering, and W.C. Moldenhauer. 1977. Conservation tillage in the eastern Corn Belt. J. Soil Water Conserv. 32:20–28.

Gupta, S.C., W.E. Larson, and D.R. Linden. 1983. Tillage and surface residue effects on soil upper boundary temperatures. Soil Sci. Soc. Am. J. 47:1212–1218.

Haas, H.J., C.E. Evans, and E.F. Miles. 1957. Nitrogen and carbon changes in Great Plains soils as influenced by cropping and soil treatments. USDA Tech. Bull. 1164. U.S. Government Printing Office, Washington, D.C.

Heichel, G.H., and D.K. Barnes. 1984. Opportunities for meeting crop nitrogen needs from symbiotic nitrogen fixation. p. 49–59. In D.F. Bezdicek (ed.) Organic farming. Spec. Pub. 46. American Society of Agronomy, Madison, WI.

Holland, E.A., and D.C. Coleman. 1987. Litter placement effects on microbial and organic matter dynamics in an agroecosystem. Ecology 68:425–433.

House, G.J., B.J. Stinner, D.A. Crossley, Jr., E.P. Odum, and G.W. Langdale. 1984. Nitrogen cycling in conventional and no-tillage agroecosystems in the Southern Piedmont. J. Soil Water Conserv. 39:194–200.

Jannsen, B.H. 1984. A simple method for calculating decomposition and accumulation of 'young' soil organic matter. Plant Soil 76:297–304.

Kitur, B.K., M.S. Smith, R.L. Blevins, and W.W. Frye. 1984. Fate of 15N-depleted ammonium nitrate applied to no-tillage and conventional tillage corn. Agron. J. 76:240–242.

Ladd, J.N. 1981. The use of ^{15}N in following organic matter turnover, with specific reference to rotation systems. Plant Soil 58:401–411.

Lal, R. 1974. Soil temperature, soil moisture and maize yield from mulched and unmulched tropical soils. Plant Soil 40:129–143.

----. 1976. No-tillage effects on soil properties under different crops in western Nigeria. Soil Sci. Soc. Am. J. 40:762–768.

Lehenbauer, P.A. 1914. Growth of maize seedlings in relation to temperature. Physiol. Res. 1:247–288.

Linn, D.M., and J.W. Doran. 1984a. Aerobic and anaerobic microbial populations in no-till and plowed soils. Soil Sci. Soc. Am. J. 48:794–799.

----, and ----. 1984b. Effect of water-filled pore space on CO_2 and N_2O production in tilled and nontilled soils. Soil Sci. Soc. Am. J. 48:1167–1272.

Lynch, J.M., and L.M. Panting. 1980. Cultivation and the soil biomass. Soil Biol. Biochem. 12:29–33.

Meisinger, J.J., V.A. Bandel, G. Stanford, and J.O. Legg. 1985. Nitrogen utilization of corn under minimal tillage and moldboard plow tillage. I. Four-year results using labeled N fertilizer on an Atlantic Coastal Plain soil. Agron. J. 77:602–611.

Mengel, D.B., D.W. Nelson, and D.M. Huber. 1982. Placement of nitrogen fertilizers for no-till and conventional till corn. Agron. J. 74:515–518.

Mielke, L.N., J.W. Doran, and K.A. Richards. 1986. Physical environment near the surface of plowed and no-tilled surface soils. Soil Tillage Res. 5: 355–366.

Parker, D.T. 1962. Decomposition in the field of buried and surface applied cornstalk residue. Soil Sci. Soc. Am. Proc. 26:559–562.

Power, J.F., J.W. Doran, and W.W. Wilhelm. 1986. Uptake of nitrogen from soil, fertilizer, and crop residues by no-till corn and soybean. Soil Sci. Soc. Am. J. 50:137–142.

––––, and R.I. Papendick. 1985. Organic sources of nutrients. p. 503–520. In O.P. Engelstad (ed.) Fertilizer technology and use, 3rd ed. Soil Science Society of America, Madison, WI.

Rice, C.W., and M.S. Smith. 1982. Denitrification in no-till and plowed soils. Soil Sci. Soc. Am. J. 46:1168–1172.

––––, and ––––. 1984. Short-term immobilization of fertilizer nitrogen at the surface of no-till and plowed soils. Soil Sci. Soc. Am. J. 48:295–297.

––––, ––––, and R.L. Blevins. 1986. Soil nitrogen availability after long-term continuous no-tillage and conventional tillage corn production. Soil Sci. Soc. Am. J. 50:1206–1210.

Rovira, A.D., and E.L. Greacen. 1957. The effect of aggregate disruption on the activity of micro-organisms in the soil. Aust. J. Agric. Res. 8:659–673.

Russel, J.C. 1939. The effects of surface cover on soil moisture losses by evaporation. Soil Sci. Soc. Am. Proc. 4:65–67.

Russel, J.S. 1960. Soil fertility changes in long-term experimental plots at Kobybolite, South Australia. I. Changes in pH, total nitrogen, organic carbon, and bulk density. Aust. J. Agric. Res. 11:902–926.

Russell, R.S., R.Q. Cannell, and M.J. Goss. 1975. Effects of direct drilling on soil conditions and root growth. Outlook Agric. 8:227–232.

Sain, P., and F.E. Broadbent. 1977. Decomposition of rice straw in soils as affected by some management factors. J. Environ. Qual. 6:96–100.

Skjemstad, J.O., R.C. Dalal, and P.F. Barron. 1986. Spectroscopic investigations of cultivation effects on organic matter of vertisols. Soil Sci. Soc. Am. J. 50:354–359.

Thomas, G.W., and W.W. Frye. 1984. Fertilization and liming. p. 87–126. In R.E. and S.H. Phillips (ed.) No-tillage agriculture. Van Nostrand Reinhold Co., New York.

Tiessen, H., J.W.B. Stewart, and J.R. Bettany. 1982. Cultivation effects on the amounts and concentration of carbon, nitrogen, and phosphorus in grassland soils. Agron. J. 74:831–835.

Tisdall, J.M., and J.M. Oades. 1982. Organic matter and water-stable aggregates in soils. J. Soil Sci. 33:141–163.

Touchton, J.J., and W.L. Hargrove. 1982. Nitrogen sources and methods of application for no-tillage corn production. Agron. J. 74:823–826.

U.S. Department of Agriculture. 1980. Report and recommendations on organic farming. U.S. Department of Agriculture, Washington, DC

Van Ouwerkerk, C., and F.R. Boone. 1970. Soil physical aspects of zero-tillage experiments. Neth. J. Agric. Sci. 18:247–261.

Van Veen, J.A., and E.A. Paul. 1981. Organic carbon dynamics in grassland soils. I. Background information and computer simulation. Can. J. Soil Sci. 61:185–201.

Varco, J.J., W.W. Frye, and M.S. Smith. 1985. Nitrogen recovery by no-till and conventional till corn from cover crops. p. 127–131. In W.L. Hargrove and F.C. Boswell (ed.) Proceedings of the 1985 Southern Region No-till Conference, Griffin, GA.

Voss, R.D., and W.D. Shrader. 1984. Rotation effects and legume sources of nitrogen for corn. p. 61–68. In D.F. Bezdicek (ed.) Organic farming. Spec. Pub. 46. American Society of Agronomy, Soil Science Society of America, and Crop Science Society of America, Madison, WI.

Wilhelm, W.W., J.W. Doran, and J.F. Power. 1986. Corn and soybean yield response to crop residue management under no-tillage production systems. Agron. J. 78:184–189.

––––, J.S. Schepers, L.N. Mielke, J.W. Doran, J.R. Ellis, and W.W. Stroup. 1985. Dryland maize development and yield resulting from tillage and nitrogen fertilization practices.

p. 27. *In* Proc., 10th Conf., Reduced Tillage-Rational Use in Sustained Production. Int. Soil Tillage Res. Org., Guelph, Ontario, Canada. 8–12 July. Elsevier Science Publishing, Amsterdam.

5 Controls on Dynamics of Soil and Fertilizer Nitrogen[1]

W. B. McGill and R. J. K. Myers[2]

Agriculture and forestry perturb cycles of energy, nutrients, and water. To understand their response to perturbations, it is essential to know the mechanisms by which nutrient, energy, or water cycles are controlled. Consequently, this chapter deals with controls. Nitrogen cycling has benefitted from recent thorough reviews (Clark and Rosswall, 1981; Stevenson, 1982). With such a background, this chapter (i) examines the thesis that system architecture is a major, but neglected, control on N dynamics; and, (ii) extends that to provide a comparative analysis of the way N supply and demand are synchronized in temperate vs. tropical agricultural systems.

Literature is cited and published examples are used to illustrate concepts. This is not an exhaustive summary of literature on the subject.

As used here, the term *architecture* relates both to the juxtaposition or physical interrelations between components of the N cycle, and to the hierarchical organization of components. The criteria separating units are size, reaction rate, or both. For example, studies of soil and fertilizer N comprise units from the colloid to broad soil-climatic zones; the range is 11 to 12 orders of magnitude (Fig. 5-1). Each component or unit, comprises smaller units below and is in turn part of a larger unit above. The challenge is to discover the links between units and apply them in understanding and managing the N cycle.

Macropedology:	Soil zone	1000 km
	Soil series	1 km
	Pedon	1 m
Micropedology:	Aggregate	1 mm
	Microorganism	1 μm
	Colloid	0.001 μm

Fig. 5-1. Hierarchy of units contributing to functioning of soil-plant systems based on size relations.

[1] Contribution of the Dep. of Soil Science, Univ. of Alberta, Edmonton, AB, T6G 2E3, Canada, and CSIRO, Div. of Tropical Crops and Pastures, St. Lucia, Queensland, Australia.

[2] Professor and chair, Dep. of Soil Science, Univ. of Alberta, Edmonton; and sub-program leader and principal research scientist, Cunningham Lab., Div. of Tropical Crops and Pastures, CSIRO, St. Lucia, Queensland, Australia, respectively.

Copyright © 1987 Soil Science Society of America and American Society of Agronomy, 677 S. Segoe Rd., Madison, WI 53711, USA. *Soil Fertility and Organic Matter as Critical Components of Production Systems,* SSSA Spec. Pub. no. 19.

Hierarchy theory may be useful. It provides a framework in which to examine interactions and links between different levels of organization. Hierarchies involve ordering of a set, the elements of which may or may not include physical objects (Webster, 1979). Soil-plant systems include the spectrum from the lowest level in Fig. 5–1 (colloid or lower) to the highest (soil zone). In general, each level both consists of and changes more slowly than the one below it. Such attributes are characteristic of both structural and dynamic hierarchies (Webster, 1979). Hence, for the present discussion the soil-plant hierarchy may be viewed as a series of concentric circles with the colloid at the center of the smallest circle; and the soil zone the largest circle.

The above view is consistent with a concept of surfaces between component elements. Allen et al. (1984) consider the problems of surfaces, of observation sets, and the potential errors and problems in linking levels of observation. The location of the observer relative to the surfaces determines which perceptions of the system are possible. Surfaces appear opaque to scientific scrutiny. Consequently, researchers at the organism level have difficulty "seeing" out to the soil environment; researchers at the pedon level can "see" organisms within only indistinctly and then only as broad groups such as "microorganisms" or "soil animals" etc. It is, therefore, dangerous to extrapolate understanding gained at a lower level to make quantitative predictions at a higher level without also making measurements at the higher level. It is equally dangerous to attempt the reverse.

Webster (1979) shows there are characteristics of each level in a hierarchy that are unique and not defined for other levels. Therefore, behavior at a higher level cannot be predicted from lower level understanding, but it is consistent with that understanding. For example, processes can be considered on at least two levels in soil. At the level of macropedology, it is possible to predict many biological reaction rates (e.g., nitrification or decomposition) on the basis of temperature, moisture, and substrate level using simple first order or Michaelis-Menten kinetics (Voroney et al., 1981; Malhi and McGill, 1982; Tanji, 1982; Myers et al., 1982; Hunt and Parton, 1986). Perhaps even fewer variables are needed in some cases. At the micropedology level, molecular diffusion; growth stage, viability, growth kinetics, energy metabolism, and environment of individual organisms; and species diversity are relevant (McLaren and Ardakani, 1972; Coleman et al., 1978; Schmidt, 1982). It would be ludicrous to attempt to create separate mathematical expressions for the activity of each individual organism and sum them up to make predictions at the macropedology level. Yet, behavior at the macropedology level must be consistent with and understood on the basis of micropedology. It is, therefore, necessary (i) to work concurrently on at least two levels; and (ii) to accept the concept of an average organism just as gas behavior is related to an average particle.

Management and productivity are concerns of this publication. Productivity may be viewed as the integrated result of land capability and management (Fig. 5–2). Land capability, however, is determined by the natural integration of soil properties with climate and plants. Soil properties in turn reflect processes operating at several levels. Management effects, while hopefully expressed at the landscape level, usually operate or influence processes at the organism, aggregate, or horizon levels. System architecture and links between levels should therefore be examined to understand controls on N cycling and to devise safe, efficient ways to use this resource.

ARCHITECTURE: MOLECULAR LEVEL

The chemistry of polymers determines their structure, which in turn controls many of their functions. For example, the three-dimensional structure of a protein may make the difference between its being an effective enzyme or a catalytically inactive protein molecule. Beyond that, however, associations of different polymers and/or monomers (architecture) generates higher order structures with unique functions. With the exception of some exudates and metabolites, most organics added to or which persist in soil are present as organism structures or as physically isolated materials. They are neither added nor stabilized independently of other molecules. For example, the cell envelop consists of identifiable molecules but their unique associations yield unified functional units (Fig. 5–3). Similarly, cellulose may be defined chemically as β-(1-4) glucose polymers. But, its properties and role in plants, and its behavior in soil, may be better understood from its architecture (Fig. 5–4). Feedback between minerals such as short-range ordered oxides which promote phenolic polymerization (Shindo and Huang, 1984) on one hand, and organics such as polycarboxylic acids which interfere with long-range crystalization of minerals (Violante and Huang, 1984) on the other, leads to higher level structures and soil components with unique properties. Behavior of such organo-mineral complexes is not predictable on the basis of assumptions that either component operates in isolation.

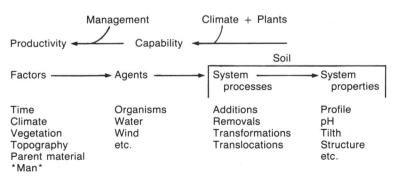

Fig. 5–2. Multilevel contributions to productivity.

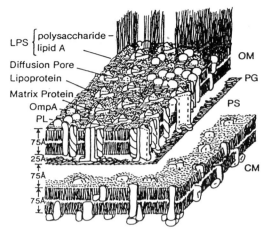

Fig. 5-3. Schematic representation of the possible molecular architecture of *E. coli* cell envelope. Abbreviations used are: PL—phospholipid, OM—outer membrane, PG—peptidoglycan, PS—periplasmic space, and CM—cytoplasmic membrane. Polysaccharide chains in only some of the LPS molecules are shown. From DiRienzo et al. (1978).

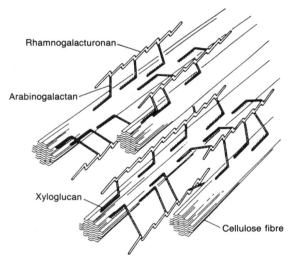

Fig. 5-4. Three-dimensional representation of the possible architecture of a primary plant cell wall showing relation of cellulose fibres to each other and their linkage with various glucans in the hemicellulose matrix. From Albersheim (1975) after Keegstra et al. (1973) with permission.

Enzymatic catalysis releases constituent monomers from polymers and is a necessary step in their decomposition and in elemental recycling from them. Such catalysis, however, is affected by associated molecules and structures. Decomposition of polymers, even though they may be well defined when considered in isolation, is not independent of architecture. Therefore, unique decomposition rates exist for unique structures (e.g., cell walls, membranes, organo-mineral complexes) rather than for groups of biochemicals (e.g., proteins) which exist in several structures or locations. Juma and McGill (1986) have further discussed relationships between architecture and organic matter dynamics.

From the standpoint of decomposition and behavior of nutrient elements in soil, it appears the architecture of molecular associations takes precedence over links between monomers, or characteristics of isolated organic or mineral soil components.

ARCHITECTURE: ORGANISM LEVEL

Soil organisms may be viewed in the short term as source-sink components of soil-plant systems and in the long term as catalysts. Immobilization of N during microbial growth on energy-rich substrates followed by remineralization constitutes short-term mineralization-immobilization turnover, or MIT as designated by Jansson and Persson (1982). Much of the remineralized N comes directly from former biomass (Paul and Juma, 1981) during its subsequent metabolism in soil (McGill et al., 1981). These events result from accumulation and depletion of biomass in localized microsites and thereby operate during time scales of days to months. Over such time intervals, changes in soil biomass must be explicitly accounted in calculating N dynamics. Over time intervals of years, the biomass of a soil reflects long-term plant activity and tends toward a steady-state characteristic of the system (McGill et al., 1986). It follows that over years and decades, soil organisms are appropriately viewed as catalysts allowing simple first-order calculations based on substrate alone.

At the organism level, substrates may also be habitats (McGill et al., 1981) and frequently comprise other organisms. Substrates and habitats must therefore be viewed at a size scale comparable to that experienced by organisms responsible for processes of interest. For example, waxes may be described on the basis of biochemical structural formulae at the molecular level. Such a view may be too fine for many purposes. They may also be perceived from the human perspective as smooth flat surface coatings. But that is about six orders of magnitude too gross. Scanning electron microscopy resolves waxes at the scale at which they are significant to soil bacteria, fungi, and even plants (Fig. 5–5). Their physical influence on decomposition, and possibly on infection, can be seen at the micrometer size scale.

Comminution of litter by soil fauna is important to nutrient cycling (Swift et al., 1979). Viewed from the human perspective, it is readily seen that slicing a cube into several parts increases surface area. Litter and plant debris, however, have characteristic internal microstructures (Fig. 5–6). Internal surface areas clearly exceed external surface area. But such surfaces are obvious only at the micrometer size scale. This is an example of the principal that measured distances around, or surface areas of,

Fig. 5–5. Wax on leaf surface of wheat (*Triticum aestivum* L.) × 2750. From Troughton and Sampson (1973) with permission.

irregular objects are inversely proportional to the size of the instrument used in their measurement. At the organism level, therefore, comminution of litter by soil fauna not only increases surface area but, perhaps more importantly, increases access to existing surfaces. Fyles and McGill (1987) concluded that differences in internal structure and surface area of several forest litters examined, influenced the rates at which they decomposed. Plant invasion by fungi through stomates rather than by dissolving the cuticle (Fig. 5–7) is another example at the organism level of the influence of system architecture on its dynamics.

Such effects are readily apparent when multiple taxa are involved. Consider for example, fungi and earthworms. An intriguing arrangement is the fungus within the earthworm (Fig. 5–8). Is the earthworm digesting plant debris or the fungus in Fig. 5–8? If it is the fungus, then the worm is a "fungal grazer" and not a primary decomposer. Alternatively, the earthworm may benefit from fungal extracellular digestion of plant debris, analogous to benefits derived by ruminants from rumen organisms. The fungus may also benefit from the moist mobile environment within the earthworm gut. Viewed from human size perspectives, earthworms appear to be primary decomposers. At the soil organism level, system architecture could conceivably be such that they are secondary decomposers. They also alter the environment of primary decomposers in such a mutualistic relationship.

Interfaces between soil abiotic components and soil organisms create a wide variety of microenvironments (Fig. 5–9 and 5–10). Such envi-

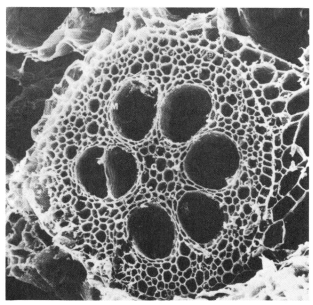

Fig. 5–6. Transverse section of the central part of a maize root (*Zea mays* L.) × 225. E—endodermis; P—pericycle; M—metaxylem; C—centre of the vascular cylinder comprising pith parenchyma cells. From Troughton and Sampson (1973) with permission.

ronmental heterogeneity provides the basis for simultaneous operation of apparently mutually exclusive processes (e.g., N mineralization or immobilization) and the concurrent existence of extremely diverse groups of soil organisms as well as microsite moisture and aeration regimes. Reviews are available on the effects of water potential on microbial ecology (Griffin, 1980) and physiology (Griffin, 1981). It is clear that the energy status of water requires compensatory adjustments by soil organisms, that the extent and mechanism of adjustment varies among taxa, and that microbial processes (spore germination, growth, and sporulation) are not all equally sensitive to water potential. Further, the energy status of soil water is related to diffusion both of solutes and gases. Most of the literature available on this subject, however, has been derived from situations in which the organism was (or was assumed to be) in contact with, or growing in, water of known potential. The arrangement of soil particles controls pore-size distribution and hence where moisture and substrates can be relative to soil organisms or enzymes. For example, soil pore radii of 0.48 and 1.89 μm would exclude 86% of the spherical and hyphal soil biovolume, respectively, based upon size distributions reported by Jenkinson et al. (1976). A radius of 0.48 μm would exclude 98% of the combined biovolume. It can also be shown that pore radii > 10 μm are normally too dry to permit microbial activity. Combining the above with pore volume distribution data of Aylmore and Sills (1978) and of Marshall and Holmes (1979), it appears that only about 40% of

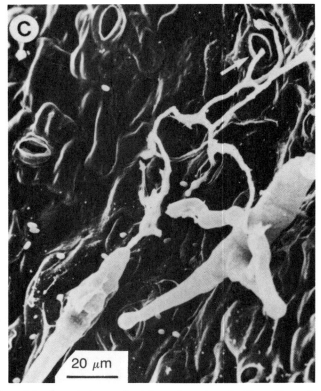

Fig. 5–7. Fungal hypha entering a plant through a stomate. From Tsuneda (1983) with permission.

the pore volume is small enough to retain water and large enough to permit entry of microbes. Moisture and substrates in the 60% that is not habitable are similarly inaccessible unless substrates can diffuse or be moved to consuming organisms. Shrinkage due to loss of water may make spaces too small even for enzyme entry. Griffin (1981) provides examples of this regarding wood decay.

Movement of soil organisms is problematic because of the control by water of their movement and the effect of soil fabric on water film thicknesses. Wong and Griffin (1976) reported that *Bacillus subtilis* moved only about 20 mm in 48 h in a soil at −5 kPa moisture potential and only about one-third as far at −10 kPa. Both moisture potentials exceed that of normal field soil conditions. Movement of bacteria should not be considered an effective mechanism to bring them into contact with fresh substrates in soil environments. Movement of larger organisms such as nematodes may be similarly restricted in soils. Wallace (1958) showed that *Heterodera schectii* moved fastest (about 100 mm h^{-1}) in water films 2 to 5 μm deep. Movement in 1-μm deep films was < 5% as fast. Water films of such thickness do not exist for long in freely drained soils.

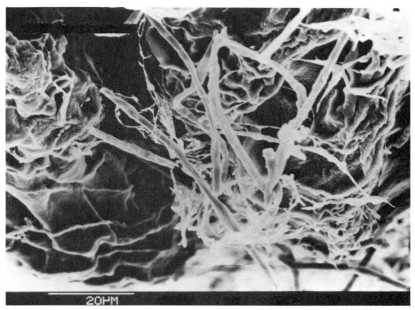

Fig. 5–8. Fungal hyphae among plant debris within the gut of an earthworm.

The tubular, hyphal form of fungi (Fig. 5–11) suits them to functioning in a soil environment where moisture is spatially heterogeneous. Survival of fungi under dry soil conditions may to a significant degree, be due to their ability to bridge dry gaps and utilize water from localized moist microsites such as interstices between soil minerals. In this situation, architectures of the organism and of its habitat are complementary. It follows that biological processes may continue due to favorable microsite environments in soils which at the macroscale are unsuited to support such activity. The environment must be described on the same size scale as the agent causing the process. Tools to do this have been lacking in the past. Electron microscopy (SEM and TEM) will be useful in this regard, as will recent developments in fiber optics and miniaturized video systems for in situ soil examinations.

Size relations among groups of soil organisms in association with pore-size distribution in soils may be a control on remineralization of immobilized N in soil. Elliott et al. (1980) hypothesized that nematodes were more restricted in their access to soil pores than are amoebae and that bacteria were least restricted. In fine soils, with small pores, amoebae should be able to enter a greater proportion of the pores than nematodes, thereby consuming a greater share of the bacterial biomass. Emergence of amoebae from small pores would make them accessible to nematodes. Consequently, nematodes should be favored by the presence of amoebae in soil microcosms and the magnitude of the effect should increase with decreasing soil-particle size. Their data, summarized in Table 5–1, are

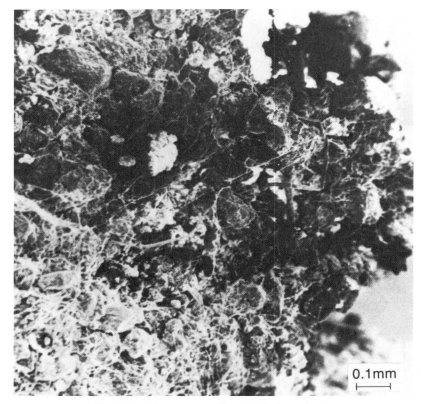

Fig. 5–9. Soil aggregate with surficial fungal hyphae.

consistent with such a hypothesis. Under such conditions, a similar beneficial effect on N mineralization would be expected if amoebae form a
link between two trophic levels; one (bacteria) in small pores and one
(nematodes) in larger pores.

Intimate associations between clay, bacteria, and plant mucigel provide a basis for higher level organization leading to formation of stable
soil aggregates (Fig. 5–12). Recent electron microscopy studies (Foster et
al., 1983) have done much to elucidate the architecture of the rhizosphere.
The TEM combined with histochemistry (Foster and Martin, 1981) now
permit some soil enzymes and biochemicals to be examined in situ. Such
observations clearly show the architectural constraints on movement both
of substrates and of products. A unique environment exists on the rhi-
zoplane (Fig. 5–13) where bacteria are abundant, and interact directly
with plant roots, but are not readily observed within the mucigel matrix.

The above examples demonstrate functional interactions between
organisms or soil fabrics of divergent sizes and morphologies. The sites
at which they occur are determined by the size of the smallest partner.

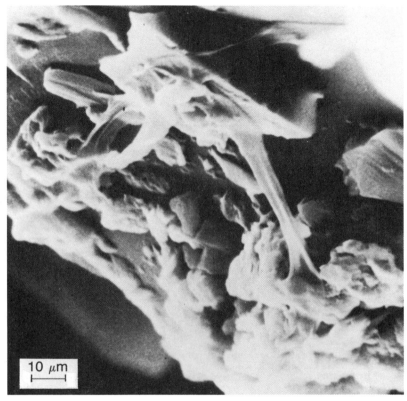

Fig. 5–10. Collapsed fungal hypha joining soil mineral particles.

The significance to the soil-plant system, however, normally must be expressed through the larger partner.

ARCHITECTURE: PEDON LEVEL

Architecture of the Ap horizon is controlled by tillage, or lack thereof. Conventional tillage reduces vertical zonation in crop residue distribution at the centimeter to decimeter size scale, although marked heterogeneity remains at the organism level. Reduced tillage permits such zonation to develop over distances of 2 to 15 cm. Do such changes in architecture at the centimeter size scale influence micrometer size organisms such as microorganisms, meter size organisms such as plants, and overall field-scale behavior toward N cycling? Present evidence suggests the answer to all three is yes. Examples of the first two are provided in this section; the latter will be discussed under management.

Carter and Rennie (1982) showed that within the profile, changing from conventional to zero tillage resulted in a reorganization of micro-

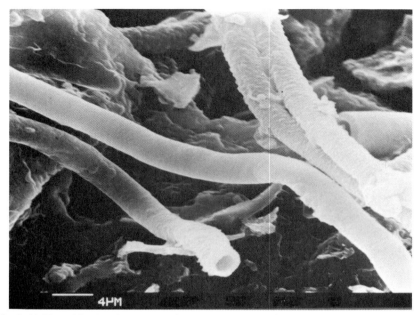

Fig. 5–11. Fungal hyphae in decomposing wood, demonstrating thin tubular character which permits them to bridge gaps between habitable sites.

Table 5–1. Proportional increase in biomass of nematodes grown with bacteria in a coarse and a fine-textured soil as a result of including amoebae. Proportional increase factor (PIF) has a value of 1.0 if nematode biomass is doubled in the presence of amoebae. Adapted from Elliott et al. (1980) with permission.

Texture	Biomass PIF†
Coarse	1.1
Fine	2.9

†Biomass PIF = (Nematodes with amoebae/nematodes without amoebae)-1.

organisms and of potential to mineralize N. Under conditions of zero tillage in a Dark Brown Chernozemic soil, net N mineralization during 12 weeks in the laboratory increased by twofold in the 0 to 2-cm layer and decreased slightly in the 2- to 4-cm layer (Table 5–2). Similar reorganization of organic residues and microorganisms has been reported by Doran (1980) in which no-tillage treatment increased organic carbon and Kjeldahl N, in the surface 0 to 7.5 cm over conventional tillage. Aerobic organisms were increased in the surface 0 to 7.5 cm and decreased in the 7.5- to 15-cm layer by no-tillage cropping. Denitrifiers increased by sevenfold in the surface layer. Doran et al. (1984) further showed that crop residues left on the surface in Nebraska maintained a higher soil moisture status, protected the surface soil from excessive heating, and altered the distribution of moisture throughout the soil profile. These influences were reflected in greater yields of maize (*Zea mays* L.) and soybean [*Glycine*

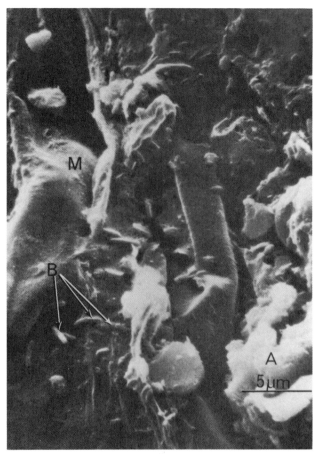

Fig. 5-12. Scanning electron micrograph of plant root mucigel (M), bacteria (B), and an aggregate (A). From Rovira et al. (1983) with permission.

max (L.) Merr.], but not sorghum [*Sorghum bicolor* (L.) Moench] in the presence of crop residues.

Spatial heterogeneity in soil conditions may influence root and mycorrhizae distribution. St. John et al. (1983) showed preferential tree root growth into litter materials rather than sandy controls at two tropical forest sites (Fig. 5-14). They also examined mycorrhizae hyphal lengths on seedling roots grown in pots with isolated organic-rich or sandy compartments. Hyphal lengths were greater in the organic materials (Fig. 5-15). This difference was attributed to favorable conditions in the organic material and not to differences in incidence of root contact or in initial inoculation. The potential therefore exists for centimeter or decimeter scale heterogeneity to be exploited by microscale (mycorrhizae) and macroscale (tree root) organisms. Because plants may selectively and effec-

Fig. 5–13. Abundance of bacteria (B) visible near a damaged portion of a root where the mucigel (M) has been stripped away. From Rovira et al. (1983) with permission.

Table 5–2. Distribution with depth of mineralizable N during a 12-week laboratory incubation of Lethbridge loam which had been managed under conventional or zero tillage systems. Adapted from Carter and Rennie (1982) with permission.

| Depth | Nitrogen mineralized during 12 weeks (Lethbridge soil) | |
	Conventional	Zero tillage
cm	mg/kg	
0–2	92	184
2–4	80	60

tively exploit favorable locations, yield may be a function of the number, location, and size of favorable sites. Bulk soil averages may not be informative in soils with such architecture. It is conceivable that crop growth

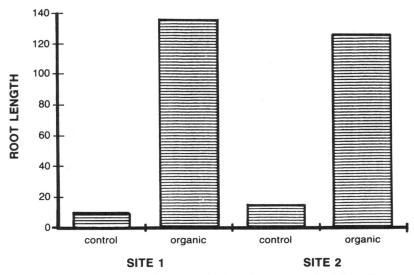

Fig. 5–14. Growth of tree roots into a sandy control or an organic rich medium at two tropical forest sites near Manous, Brazil. Adapted from St. John et al. (1983) with permission.

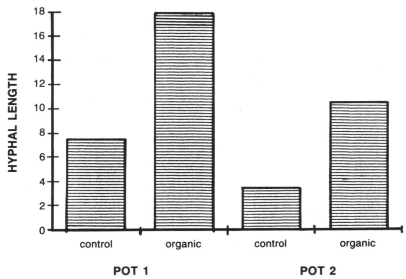

Fig. 5–15. Lengths (dm) of hyphae of *Glomus fasiculatum* inoculated onto seedlings of *Fragaria* spp. Hyphal lengths were measured in separated sandy and organic compartments of pots maintained in a growth chamber. Adapted from St. John et al. (1983) with permission.

may be greater than expected if nutrients are concentrated where plant roots have a competitive advantage over other organisms or reactive soil components. Alternatively, architectures in which essential nutrients are trapped in inaccessible environments, or in areas where plant root activity is restricted (e.g., dry zones) may result in unexpectedly low yields.

Water movement at the pedon level is controlled by both soil horizon and aggregate level properties. Horizon level behavior may result from frozen layers, textural changes, mineralogical changes, and macrostructure. Aggregate level behaviors result from the distribution of micro- and macropores with their associated retained and mobile phases, respectively (Addiscott et al., 1978).

Solutes in the mobile phase move with mobile water, while they are retained within aggregates by the nonmobile water phase. Solutes in the nonmobile phase equilibrate with the mobile phase through diffusion. Movement of solute from within aggregates in structured soils thereby exhibits a characteristic skewed leaching pattern caused by aggregation. At the horizon level, vertical cracks can permit water to bypass large volumes of soil, thereby not moving solutes. Alternatively, solutes already present in the water move further than would be expected. Kissel et al. (1974) observed that large unconnected pores in a swelling clay soil at Temple, TX enhanced nitrate (NO_3^-) leaching by allowing percolating rainwater to bypass water contained within other soil structural units. Similarly, White et al. (1986) examined leaching of bromacil (5-bromo-3-sec-butyl-6-methyluracil) and napropamide [2-(a-naphthoxy)-N,N-diethylpropionamide] through a structured Evesham clay. Water flowed preferentially down natural cracks and worm channels. The herbicides were added to dry soil at the start of leaching. With continuous leaching (12 mm h^{-1}) 85% of the added napropamide and nearly all the bromacil were leached out of the soil by one pore volume of water. Swelling clays may, however, upon expansion slow water movement, and increase equilibration between mobile and nonmobile phases thereby increasing leaching of pesticides from clay soils (Majka and Lavy, 1977).

Either complementary or opposing behaviors may be observed at the pedon level as a result of interactions of these lower level controls on water movement. Work on movement of tritiated water during spring thaw of structured soils (Miller and McGill, 1983, unpublished data, University of Alberta) provides some insight into these behaviors. A finite difference model of water movement in a Gray Luvisolic soil was developed. Tritiated water was added to the soil in December, and its movement together with changes in water content were measured during spring thaw. Model results show the skewed solute pattern typical of water movement in structured soils where solute can be held back in nonmobile water within aggregates and shot forward in the mobile phase. Frozen layers, however, impede water movement, increase degree of equilibration between mobile and nonmobile phases, and result in solute behavior in thawing structured soil which is identical to behavior of solutes in unfrozen unstructured soils (Fig. 5–16). This is an example of pedon level

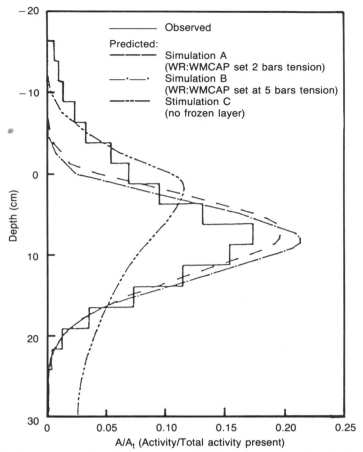

Fig. 5–16. Comparison of simulated patterns of ^3H_2O movement in a Gray Luvisolic soil at Breton, Alberta during thawing with observed distribution of ^3H_2O. WR:WMCAP determines the retained vs. mobile segregation of water in the simulations used; with all water held at tensions less than the assigned value, permitted to move in the mobile phase. A/At is the ratio of tritium activity in a layer to total tritium activity (J.C. Miller, M.Sc. thesis, 1983. University of Alberta).

behavior being sensitive to controls at the horizon level that mask behaviors at the aggregate level.

Furthermore, water movement and associated solute leaching through the pedon are subject to feedback between moisture, nutrient supply, and crop growth. Data of Campbell et al. (1984), for crop rotation studies on a Brown Chernozemic soil illustrate this effect. They observed greater accumulations of NO_3^- below 90 cm in soils receiving no added fertilizer N than in those fertilized with N (Fig. 5–17). They also observed less NO_3^- below 90 cm where a deep rooting forage was included in the rotation. Such data suggest that assuring rooting activity through fertilizers or species selection, increases either uptake of previously leached NO_3^-

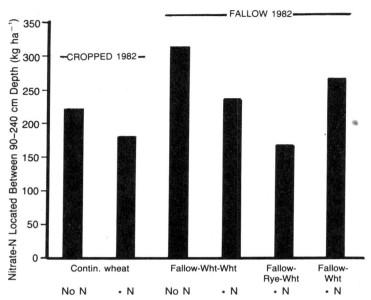

Fig. 5–17. Accumulation of NO_3^--N below 90 cm in a Brown Chernozemic soil at Swift Current Saskatchewan following 15 yr under diverse cropping and fertilizer regimes. From Campbell et al. (1984) with permission.

or reduces leaching of NO_3^- during the current crop year. A major control on NO_3^- accumulation is therefore rooting activity. Most concepts of NO_3^- leaching include a linear link between NO_3^- added and leaching. It is also widely assumed that NO_3^- leaching is more prevalent in mesic or humid soil environments than in semiarid soil environments. The cybernetic nature of soil-plant systems with associated feedback between moisture, root activity, and nutrients denies the generality of such linear logic.

Further research is warranted on the cybernetics of pedon level architecture, spatial distribution of plant-root activity, and crop growth. Root activity as a function of spatially heterogeneous resources or management practices such as fertilization or irrigation may be a major determinant of fertilizer-use efficiency, and losses of added nutrients.

MANAGEMENT: FIELD-SCALE ARCHITECTURE

Data of Malhi and Nyborg (1985) provide an example of field-scale expression of management effects which operate at the organism and aggregate levels. They have shown that altering the architecture of the soil-fertilizer-plant system influences fertilizer uptake by and yield of barley (*Hordeum vulgare* L.) (Table 5–3). While fall-applied fertilizer N was not recovered by the crop as efficiently as was spring-applied N, there

Table 5–3. Barley yield and N recovery using band or nest urea placement in a Black Chernozemic soil from Alberta in the fall compared to band placement in the spring. Adapted from Malhi and Nyborg (1985), data for Site 10, with permission.

| | Fall | | | Spring | |
	Control	Band	Nest	Band	LSD
Yield, Mg/ha	2.4	3.3	3.6	3.9	0.4
Nitrogen recovered (Grain + straw), %		41	50	62	8.4

were no differences in yield between fall-nested and spring-banded N. In bands, added N is exposed to greater contact with soil than is the case in nests where N is concentrated in smaller zones. Consequently, nests reduce the potential for nitrification, immobilization and, if the soil becomes anaerobic, denitrification (Monreal et al., 1986). Nitrogen placed in nests in the fall was recovered in the crop the following year to a significantly greater extent than banded N which is more accessible at the microscale to soil organisms. Their data suggest technologies of altering system architecture to conserve nutrient resources without reliance on chemical inhibitors etc. Such technologies use physical rather than chemical techniques to control soil processes and thereby manage nutrient dynamics. If proven generally effective they could yield major economic benefits.

Tillage appears also to modify nutrient dynamics. The effect of tillage on system architecture at the pedon level has been demonstrated. Effects on field-scale dynamics over several years are apparent in data of Meisinger et al. (1985). Changes in pedon level architecture associated with changes in tillage would be expected to alter dynamics of N. Accumulation of crop residues at the surface increases the potential for N immobilization and also denitrification in that zone. Immobilized N would be expected to be remineralized over time, but in relocated biological cycling systems (high biomass and activity in the surface residues) developed in response to altered architecture. At some point, remineralization should equal or exceed immobilization. Consequently, crop N demand could be met increasingly from previously immobilized N. Such N must of course be present at a place and time when plant roots can use it. Figure 5–18 shows this trend over a 4-yr period of maize production on plow- and zero-tilled plots in Maryland. Initially, when converted from conventional plow tillage, the zero-tilled system required more N to produce comparable yields to the plow-tilled plots. That difference tended to diminish with time such that after 4-yr the two tillage systems produced comparable maize yields with comparable N additions. The zero-tilled plot may have been slightly more responsive to added N than the plowed plot.

Changes in pedon level architecture which affect organism level function are thereby seen to be expressed at the landscape or field level by changes in behavior over time. It follows that management studies in-

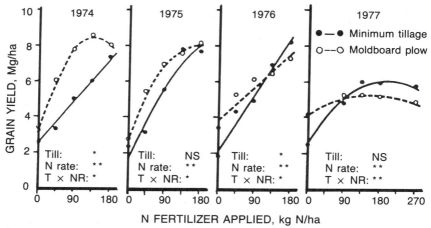

Fig. 5–18. Maize yield over a 4-yr period on a Mattapex silt loam in Maryland as affected by added N under two tillage regime. Adapted from Meisinger et al. (1985) with permission.

volving such architectural changes must also incorporate time course components. According to the analysis of Allen et al. (1984), such dynamics require a change of extent in the observation set. Such changes in dynamics lead to the question of synchronization of plant and soil processes in time as well as in space.

SYNCHRONIZATION: SUPPLY AND DEMAND

Efficient use of moisture and nutrients by crops requires that two conditions be met: i) plants must be growing when these resources are most available, and ii) plant root activity must be greatest in locations where resources are most available. Although (ii) has received greatest attention, (i) is also important.

The world's dryland agricultural climates range from those that are almost continually warm and humid (e.g., the humid tropics) to those that are moist and warm for only part of the year (e.g., cool temperate). According to where they lie within this range, the number of crops grown per year varies from one (moisture or temperature-limited) to as many as three. Timing of crop plantings depends on many factors, and it is important to examine how crop cycles fit in with N mineralization in the soil.

This issue will be examined by comparing crop uptake of N and N mineralization in the field under temperate conditions. A temperature × moisture index (M×T: a multiplicative index of relative rate of soil biological activity based upon monthly temperature and rainfall data) will then be used to compare plant-soil synchronization over a range of climatic zones. Such an index is not applicable to detailed simulation modelling but is useful for comparisons between regions with different climates.

Figure 5–19 presents results of studies at two sites in southern Manitoba. The course of N uptake by wheat (*Triticum aestivum* L.) was compared with net N mineralization on four and six plots, respectively, replicated four times. Although the two processes are reasonably synchronized, plant uptake of N tends to precede maximum N mineralization rate. Consequently, N present at seeding time has a significant bearing on total plant uptake of N. The two processes are closely correlated suggesting the potential for concurrent control of both soil and plant processes.

Similarly, a moisture × temperature index for Edmonton with a cool temperate climate shows close corerspondence of plant and soil processes (Fig. 5–20). This is consistent with the observations from southern Manitoba where temperatures are somewhat warmer. Such relationships hold for the U.S. Corn Belt (as represented by Des Moines, IA) where only one crop is grown each year (Fig. 5–20).

In the humid tropics (e.g., Tanjong Karang, Indonesia), conditions for N mineralization (as reflected by the M×T index) are high for half the year, declining only as rainfall intensity declines in the so-called dry season (Fig. 5–21). Between crops there is some opportunity for mineral N accumulation. The third crop in the cycle (July–October) is likely to be less well supplied with N than the other two crops; a factor that is consistent with the local practice of growing a legume—either groundnut or rice bean—at this time.

In the semiarid tropics (e.g., Hyderabad, India) although temperature is high throughout the year, N mineralization is seasonal in response to wet-dry seasonal conditions (Fig. 5–21). Thus, N mineralization (as re-

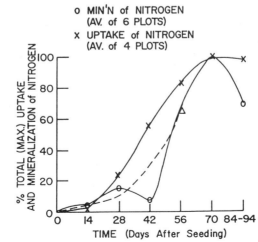

Fig. 5–19. Comparison of course of net N mineralization and quantity of N in aboveground plant parts for spring seeded wheat in southern Manitoba (W.B. McGill, M.Sc. thesis. 1969. University of Manitoba).

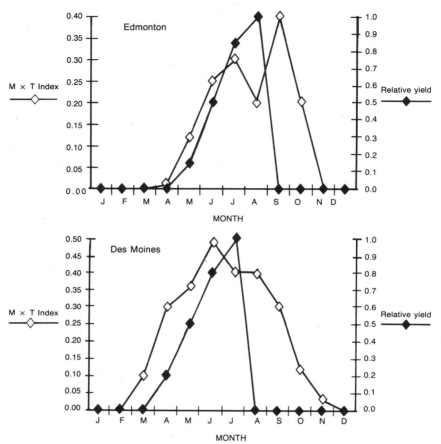

Fig. 5–20. Moisture × temperature index compared to relative crop growth rate (yield) for two sites having temperate climatic conditions.

flected by the M×T index) is likely to be rapid early in the wet season, supplying both the first (kharif) crop, and the second (rabi) crop NO_3^- also. From the viewpoint of N, the rabi crop should be fertilized, but because of dry soil conditions, N fertilizer is poorly utilized. Grain legume might be better.

In some environments, even worse asynchrony of N supply and N requirement may occur. Thus, for summer-grown grain sorghum on the Darling Downs of Queensland, Australia, crop N requirements peak between February and April, whereas supply of N in the soil would be expected to peak between October and January (Fig. 5–22). Wheat sown in early winter with peak N requirement between August and September similarly will rely on residual N accumulated during the summer when the land is fallow (Fig. 5–22). In both cases, a major reason for the mineralization index being low is the depletion of soil water by the crop.

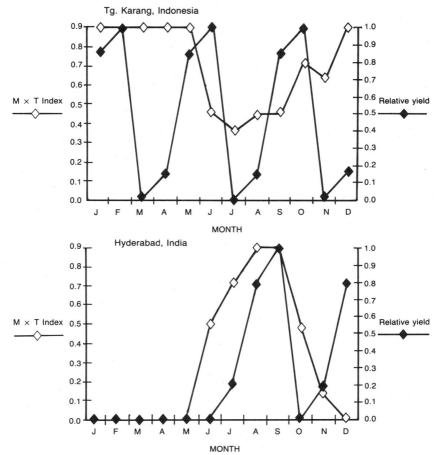

Fig. 5–21. Moisture × temperature index compared to relative crop growth rate (yield) for two of tropical sites; Tanjong Karang in the humid tropics, Hyderabad in the semiarid tropics.

Comparing the two, it appears that similar quantities would be mineralized annually with each cropping system.

The above analysis suggests most cropping systems are synchronized with potential soil biological activity. Noteable exceptions are the production of summer-grown grain sorghum or winter-grown wheat in the Darling Downs area of Australia. The potential exists, therefore, for large losses of N under such conditions. They may exist in large areas of dryland cropping in subtropical and also in semiarid tropical environments.

There is a need to devise cropping and nutrient management systems in which the timing of N release from decomposing plant residues is synchronized with crop demand for N. Such systems could be further extended to utilize N immobilization by plant residues followed by re-

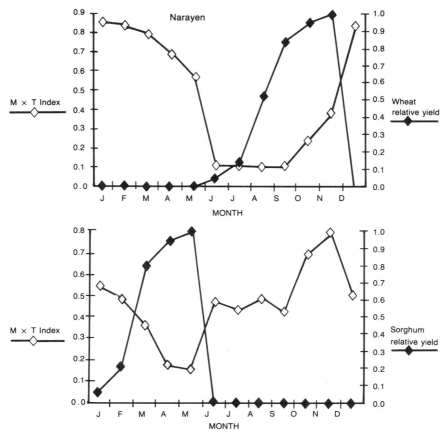

Fig. 5–22. Moisture × temperature index compared to relative crop growth rate (yield) for two crops on the same site (subtropical) in the Darling Down of Queensland.

mineralization during the period of crop growth as a means to protect fertilizer N from leaching and denitrification, yet making it available to crops. Development of such strategies, however, requires understanding of mechanisms controlling N release from both microbial biomass and plant residues and the influence on them of annual cycles of moisture and temperature.

SUMMARY

An approach to understanding soil and fertilizer N dynamics is presented that emphasizes: i) architectural relations within soil-plant systems and ii) timing of soil biological activities and crop requirements. The aim is to use such a paradigm to guide development of safe efficient use of N.

From the standpoint of decomposition and behavior of nutrient elements in soil, it appears the architecture of molecular associations takes precedence over links between monomers or characteristics of isolated organic or mineral soil components. Considerations of residue architecture will permit decomposition to be understood more clearly and modelled more explicitly.

Dynamics of N are controlled by interactions between soil abiotic components, soil organisms, and plants under control of their environment. The environment must be described on the same size scale as the agent causing a given process. The sites at which interactions occur are determined by the size of the smaller partner; the significance to the soil-plant system, however, is normally expressed through the larger partner.

Effects of system architecture are seen all the way up to crop yield at the landscape level. Further research is warranted on the cybernetics of pedon level architecture, spatial distribution of plant root activity, and crop growth. Such research should include field techniques such as tillage and nutrient placement, and their interaction both with plant root activity and with the soil environment to improve nutrient and water-use efficiency. The possibility exists to extend such concepts to nutrient and water use or allocation in intercropping systems.

The balance between concurrent mineralization and immobilization of N controls the supply of mineral N to plants, and is in turn sensitive to plant residue placement. It is suggested that the time at which mineral N (from fertilizers or soil organic matter) is made available to crops may be more closely synchronized with crop demand by manipulating mineralization/immobilization relations. The potential exists through use of crop residues to immobilize added fertilizer N followed by remineralization during times of crop demand, to prevent leaching and denitrification losses. Such technologies would require additional information on mechanisms controlling N release from both microbial biomass and plant residues and the influence on them of annual cycles of moisture and temperature.

ACKNOWLEDGMENTS

We thank N.G. Juma and G.A. Peterson for constructive comments and assistance. The assistance of M. Goh in preparation of the manuscript is gratefully acknowledged.

REFERENCES

Addiscott, T.M., D.A. Rose, and J. Bolton. 1979. Chloride leaching in the Rothamsted drain gauges: Influence of rainfall pattern and soil structure. J. Soil Sci. 29:305–314.

Albersheim, P. 1975. The wall of growing plant cells. Sci. Am. 232(4):80–95.

Allen, T.F.H., R.V. O'Neill, and T.W. Hoekstra. 1984. Interlevel relations in ecological research and management: Some working principles from hierarchy theory. USDA-FS, General Tech. Rep. RM-110.

Aylmore, L.A.G., and I.D. Sills. 1978. Pore structure and mechanical strength of soils in relation to their constitution. p. 69–78. *In* W.W. Emerson et al. (ed.) Modification of soil structure. John Wiley and Sons, New York.

Campbell, C.A., R. De Jong, and R.P. Zentner. 1984. Effect of cropping, summerfallow and fertilizer nitrogen on nitrate-nitrogen lost by leaching on a Brown Chernozemic soil. Can. J. Soil Sci. 64:61–74.

Carter, M.R., and D.A. Rennie. 1982. Changes in soil quality under zero tillage farming systems: Distribution of microbial biomass and mineralizable C and N potentials. Can. J. Soil Sci. 62:587–597.

Clark, F.E., and T. Rosswall (ed.) 1981. Terrestrial nitrogen cycles: Processes, ecosystem strategies and management impacts. *In* Ecological Bulletins (Stockholm) 33. Proc. Int. Workshop. 16–22 Sept. 1979. SCOPE/UNEP Int. Nitrogen Unit Royal Swedish Academy Sci. Commission Res. Natural Resources Swedish Counc. Planning Coord. Res., Gysinge Värdshus, Österfärnebo, Sweden.

Coleman, D.C., C.V. Cole, H.W. Hunt, and D.H. Klein. 1978. Trophic interactions in soils as they affect energy and nutrient dynamics: I. Introduction. Microb. Ecol. 4:345–349.

Di Rienzo, J.M., K. Nakamura, and M. Inouge. 1978. The outer membrane proteins of gram-negative bacteria: Biosynthesis, assembly and functions. Ann. Rev. Biochem. 47:481–532.

Doran, J.W. 1980. Soil microbial and biochemical changes associated with reduced tillage. Soil Sci. Soc. Am. J. 44:765–771.

————, W.W. Wilhelm, and J.F. Power. 1984. Crop residue removal and soil productivity with no-till crop, sorghum, and soybean. Soil Sci. Soc. Am. J., 48:640–645.

Elliott, E.T., R.V. Anderson, D.C. Coleman and C.V. Cole. 1980. Habitable pore space and microbial trophic interactions. Oikos 35:327–335.

Foster, R.C., and J.K. Martin. 1981. *In situ* analysis of soil components of biological origin. p. 75–112. *In* E.A. Paul and J.N. Ladd (ed.) Soil biochemistry, Vol. 5. Marcel Dekker, New York.

Foster, R.C., A.D. Rovira, and T.W. Cock. 1983. Ultrastructure of the root-soil interface. American Phytopathological Society, St. Paul.

Fyles, J.W., and W.B. McGill. 1987. Decomposition of boreal forest litters from central Alberta under laboratory conditions. Can J. For. Res. 17:109–114.

Griffin, D.M. 1980. Water potential as a selective factor in the microbial ecology of soils. p. 141–151. *In* J.F. Parr et al. (ed.) Water potential relations in soil microbiology. Spec. Pub. 9. Soil Science Society of America, Madison, WI.

————. 1981. Water and microbial stress. p. 91–131. *In* M. Alexander (ed.) Advances in microbial ecology, Vol. 5. Plenum Publishing Corp., New York.

Hunt, H.W., and W.J. Parton. 1986. The role of modelling in research on microfloral and faunal interactions in natural and agroecosystems. p. 443–494. *In* M.J. Mitchell and J.P. Nakas (ed.) Microfloral and faunal interactions in natural and agro-ecosystems. Nijhoff/Junk, Boston.

Jansson, S.L., and J. Persson. 1982. Mineralization and immobilization of soil nitrogen. *In* F.J. Stevenson (ed.) Nitrogen in agricultural soils. Agronomy 22:229–252.

Jenkinson, D.S., D.S. Powlson, and R.W.M. Wedderburn. 1976. The effects of biocidal treatments on metabolism in soil—III. The relationship between soil biovolume measured by optical microscopy, and the flush of decomposition caused by fumigation. Soil Biol. Biochem. 18:189–202.

Juma, N.G., and W.B. McGill. 1986. Decomposition and nutrient cycling in agro-ecosystems. p. 74–136. *In* M.J. Mitchell and J.P. Nakas (ed.) Microfloral and faunal interactions in natural and agro-ecosystems. Nijhoff/Junk, Boston.

Kissel, D.E., J.T. Ritchie, and E. Burnett. 1974. Nitrate and chloride leaching in a swelling clay soil. J. Environ. Qual. 3:401–404.

Majka, J.T., and J.L. Lavy. 1977. Adsorption, mobility, and degradation of cyanazine and diuron in soils. Weed Sci. 25:401–406.

Malhi, S.S., and W.B. McGill. 1982. Nitrification in three Alberta soils: Effect of temperature, moisture and substrate concentration. Soil Biol. Biochem. 14:393–399.

————, and M. Nyborg. 1985. Methods of placement for increasing the efficiency of N fertilizers applied in the fall. Agron. J. 77:27–32.

Marshall, T.J., and J.W. Holmes. 1979. Soil physics. Cambridge University Press, London.

McGill, W.B., K.R. Cannon, J.A. Robertson, and F.D. Cook. 1986. Dynamics of soil microbial biomass and water soluble organic C in Breton L after 50 years cropping to two rotations. Can. J. Soil Sci. 66:1–19.

————, H.W. Hunt, R.J. Woodmansee, and J.O. Reuss. 1981. Phoenix: A model of the dynamics of carbon and nitrogen in grassland soils. *In* F.E. Clark and T. Rosswall (ed.) Terrestrial nitrogen cycles: Processes, ecosystem strategies, and management impacts. Ecol. Bull. 33:49–116.

McLaren, A.D., and M.S. Ardakani. 1972. Competition between species during nitrification in soil. Soil Sci. Soc. Am. Proc. 36:602–606.

Meisinger, J.J., V.A. Bandel, G. Stanford, and J.O. Legg. 1985. Nitrogen utilizaton of corn under minimal tillage and moldboard plow tillage. I. Four year results using labeled N fertilizer on an Atlantic coastal plain soil. Agron. J. 77:602–611.

Monreal, C.H., W.B. McGill, and M. Nyborg. 1986. Spatial heterogeneity of substrates: Effects on hydrolysis, immobilization and nitrification of urea-N. Can. J. Soil Sci. 66:499–511.

Myers, R.J.K., C.A. Campbell, and K.L. Weier. 1982. Quantitative relationship between net nitrogen mineralization and moisture content of soils. Can. J. Soil Sci. 62:111–124.

Paul, E.A., and N.G. Juma. 1981. Mineralization and immobilization of soil nitrogen by microorganisms. *In* F.E. Clark and T. Rosswall (ed.) Terrestrial nitrogen cycles: Processes, ecosystem strategies and management impacts. Ecol. Bull. 33:179–194.

Rovira, A.D., G.D. Bowen, and R.C. Foster. 1983. The significance of rhizosphere microflora and mycorrhizae in plant nutrition. p. 61–93. *In* A. Läuchli and R.L. Bieleski (ed.) Encyclopedia of plant physiology, new series, Vol. 15. Springer-Verlag New York, New York.

Schmidt, E.L. 1982. Nitrification in soil. *In* F.J. Stevenson (ed.) Nitrogen in agricultural soils. Agronomy 22:253–288.

Shindo, H., and P.M. Huang. 1984. Catalytic effects of manganese (IV), iron (III), aluminum, and silicon oxides on the formation of phenolic polymers. Soil Sci. Soc. Am. J. 48:927–934.

Stevenson, F.J. (ed.) Nitrogen in agricultural soils. Agronomy 22.

St. John, T.V., D.C. Coleman, and C.P.P. Reid. 1983. Growth and spatial distribution of nutrient-absorbing organs: Selective exploitation of soil heterogeneity. Plant Soil 71:487–493.

Swift, M.J., O.W. Heal, and J.M. Anderson. 1979. Decomposition in terrestrial ecosystems. University of California Press, Berkeley.

Tanji, K.K. 1982. Modelling of the soil nitrogen cycle. p. 721–772. *In* F.J. Stevenson (ed.) Nitrogen in agricultural soils. Agronomy 22:721–772.

Troughton, J.H., and F.B. Sampson. 1973. Plants: A scanning electron microscope survey. John Wiley and Sons, New York.

Tsuneda, A. 1983. Fungal morphology and ecology. Tottori Mycological Institute, Kokoge, Japan.

Violante, A., and P.M. Huang. 1984. Nature and properties of pseudoboehmites formed in the presence of organic and inorganic ligands. Soil Sci. Soc. Am. J. 48:1193–1201.

Voroney, R.P., J.A. Van Veen, and E.A. Paul. 1981. Organic C dynamics in grassland soils. 2:Model validation and simulation of the long-term effects of cultivation and rainfall erosion. Can. J. Soil Sci. 61:211–224.

Wallace, H.R. 1958. Movement of eelworms: I. The influence of pore size and moisture content of the soil on the migration of larvae of the beet eelworm *Heterodera schactii* Schmidt. Ann. Appl. Biol. 46:74–85.

Webster, J.R. 1979. Hierarchical organization of ecosystems. p. 119–129. *In* E. Halfon (ed.) Theoretical systems ecology. Academic Press, New York.

White, R.E., J.S. Dyson, Z. Gerstland, and B. Yaron. 1986. Leaching of herbicides through undisturbed cores of a structured clay soil. Soil Sci. Soc. Am. J. 50:277–283.

Wong, P.T.W., and D.M. Griffin. 1976. Bacterial movement at high matric potentials. I. *In* Artificial and natural soils. Soil Biol. Biochem. 8:215–218.

6 Controls on Dynamics of Soil and Fertilizer Phosphorus and Sulfur[1]

J. W. B. Stewart and A. N. Sharpley[2]

In this chapter the processes involved in the distribution of P and S in soils and how these processes affect the bioavailability of these elements in different soils and climates are reviewed. Renewed interest in this topic comes, in part, from actual and proposed changes in soil management under conservation tillage in temperate climates (Gebhardt et al., 1985) and from the thrust to improve the productivity of more weathered soils in the tropics (Sanchez et al., 1982). Much has already been gained from examination of the forms and distribution of P and S in natural ecosystems (Walker and Syers, 1976; Freney and Williams, 1983; Smeck, 1973, 1985). The balance between these forms in soil can be drastically altered by introduction of agriculture or changes in cropping practices. This can result in changes in the amount and distribution of C in the soil, as well as changing physical factors such as soil moisture and temperature, which control abiotic and biotic processes. Hypotheses on C, N, S, and P interaction during decomposition processes in soil have been developed. These concepts, when integrated with environmental factors, give a logical framework to observe changes in soil organic matter quality across environmental gradients and in chronology of cultivation sequences (Stewart, 1984). Together with the development of more precise methods of measuring the quantities and forms of P and S in soil, particularly microbial forms (Hedley and Stewart, 1982) we are now in a position to reexamine P and S cycling in soils and the changes in soil fertility due to farming practices.

PHOSPHORUS

Conceptual P Cycle

Figure 6–1 presents a conceptual P cycle with its components and measurable fractions in which the following processes are depicted. Primary P minerals are slowly dissolved during weathering processes pro-

[1] Saskatchewan Institute of Pedology Pub. R465, Saskatoon, Saskatchewan, SK Canada.
[2] Professor of soil science, Univ. of Saskatchewan, Saskatoon, SK Canada; soil scientist, USDA-ARS, Durant, OK 74702, respectively.

Copyright © 1987 Soil Science Society of America and American Society of Agronomy, 677 S. Segoe Rd., Madison, WI 53711, USA. *Soil Fertility and Organic Matter as Critical Components of Production Systems,* SSSA Spec. Pub. no. 19.

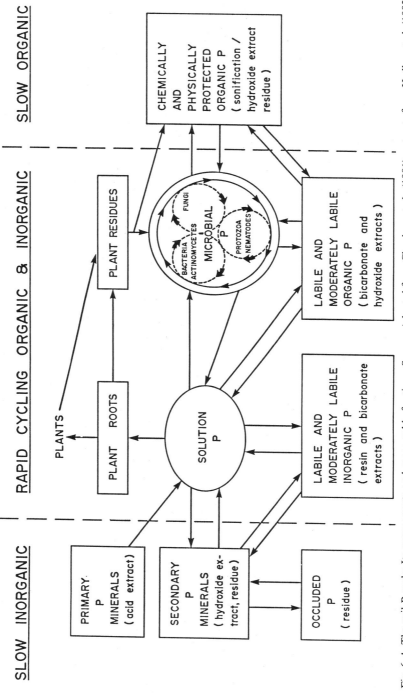

Fig. 6–1. The soil P cycle: Its components and measurable fractions. Source: Adapted from Chauhan et al., (1981); extracts from Hedley et al. (1982).

mary P minerals are slowly dissolved during weathering processes providing phosphate ions that enter into the solution P pool (cf. reviews Cole et al., 1977; Smeck, 1985; Stewart and McKercher, 1983; Tate, 1984; Tiessen and Stewart, 1985). Soil solution phosphate ions are shown to be in equilibrium with a quantity of labile inorganic P (P_i), such that in any one soil the ratio of labile inorganic P to solution P maintains a constant ratio over the normal range of P concentration found in cultivated soils. This ratio is called the *capacity factor*. Size of the capacity factor depends on physical and chemical properties of the adsorbing colloids and has been measured from < 10 to > 1000. A portion of the solution P will be precipitated as secondary P minerals and eventually converted to occluded (unavailable) forms in more weathered soils. The heterogeneity of the soil system and the large number of P reaction products makes analysis of the different P mineral species difficult and indirect approaches are usually taken to try to identify key compounds that partially control P in solution (Lindsay and Vlek, 1977; Williams et al., 1971).

Plants appear to take up exclusively phosphate ions from the soil solution and the dynamics of P uptake are well researched (Barber, 1984). The depleted solution P pool is immediately replenished from labile and moderately labile P_i forms. If these pools are depleted less soluble species such as secondary P minerals determine the solution P concentration in the soil.

Uptake of solution phosphate by bacteria and fungi, stimulated by the addition of microbial substrates, such as litter and crop residues, and release of both solution phosphate ions (P_i) and labile and stable organic (P_o) forms, as the result of cell lysis or predation (Coleman et al., 1983; Tiessen et al., 1984b), are represented in diagram (Fig. 6–1) as a revolving wheel. This is done deliberately to emphasize the central role of the microbial population in P cycling. Within the microbial cell P exists as a wide variety of compounds principally RNA (30–50%), acid soluble P_i and P_o (15–20%), phospholipids ($<10\%$) and DNA (5–10%) (Stewart and McKercher, 1982). If the microbial cell is ruptured or lysed, all these compounds will be released to the soil solution to react with both inorganic and organic soil components to form a host of P_i and P_o compounds of differing solubility or susceptibility to mineralization. The rate of mineralization of P_o forms depends largely on phosphatase activity which, in turn, can be controlled by solution P concentration (McGill and Cole, 1981). Stable P_o accumulates in both chemically resistant and aggregate protected forms (Marshall, 1971; Tisdall and Oades, 1982). The long-term changes in soil P_o have been well documented (Haas et al., 1961; Sharpley and Smith, 1983), but few attempts have been made to measure the dynamics of short-term processes that cause the long-term changes in soil P (Cole et al., 1978; Chauhan et al., 1979, 1981).

Organic phosphorus existing in chemically or physically protected forms may be slowly mineralized as a by-product of overall soil organic matter mineralization or by specific enzyme action in response to the need for P. Therefore, organic matter turnover as well as solution P_i

concentration and the demand for P by microbial and plant components will be factors controlling the lability of P_o (McGill and Cole, 1981). A continuous drain on soil P pools by cultivation and crop removal will rapidly deplete labile P_i and P_o forms (Hedley et al., 1982; Sharpley, 1985; Sharpley and Smith, 1985).

Figure 6–1, therefore, attempts to depict the P cycle as a system in dynamic equilibrium with interchanges governed by chemical, physical, and biological reactions. Microbial activity is depicted as a wheel rotating in the soil in response to energy (C) inputs and having a central role in P transformations. Should the wheel be stopped or slowed down by lack of C inputs or by partial soil sterilization, the supply of P to plants will be limited to the quantity of labile P_i. If the wheel is operating then the plant has a much larger quantity of labile P as solution P is constantly being replenished from labile P_i and P_o forms.

Phosphorus Transformation During Pedogenesis

During soil development, P composition in the soil will change (Walker and Syers, 1976). Using the measurable components of this conceptual diagram (Fig. 6–1), Tiessen et al. (1984a) showed the distribution of P across the different fractions and their relationships to other soil chemical properties in 168 USDA-SCS benchmark soils, representing eight soil orders of Soil Taxonomy. Correlation and regression analysis of P distribution and chemical analyses confirmed the partial dependence of organic matter accumulation on available forms of P. Weathering indicators such as base saturation were related to secondary P forms. Path analysis of relationships between labile and stable P forms showed that in Mollisols most of the labile resin extractable P (86%) was derived from labile inorganic forms (Fig. 6–2). In more weathered Ultisols, 80% of the variability in labile P_i was accounted for by organic forms of P (Fig. 6–3), suggesting that mineralization of P_o may be a major determinant of P fertility in these soils. The transformations of P_i and P_o in soil are therefore closely interrelated, since P_i is a source of P uptake for both plants and soil organisms and P_o may replenish solution P_i through hydrolysis, or be stabilized with the mineral phase of the soil.

This should be of particular importance in extremely weathered soils with high active Al and Fe contents, where P ions are rapidly chemisorbed and removed from solution. Organic phosphorus forms are, therefore, important and the immobilization-mineralization process becomes an important mechanism of conserving P in the soil-plant system.

Supply of Plant-Available Phosphorus

In most soils, fertilizer P application is needed to supplement native soil P_i and maintain crop demands for P_i and produce required yields. The P_i has generally been considered the major source of available P for plant uptake in soils from temperate regions (Russell, 1973). Conse-

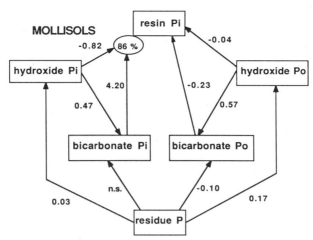

Fig. 6–2. Relationships between labile and stable components of the soil P model, shown as nonnormalized regression coefficients of a path analysis in Mollisol A horizons (% figures indicate the partial correlation between connected fractions. Variability in the labile resin P_i pool is explained to 86% by variations in the moderately available hydroxide and bicarbonate P_i fractions). Source: Tiessen et al. (1984a).

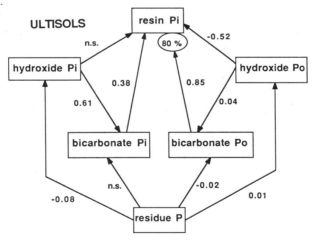

Fig. 6–3. Relationships between labile and stable components of the soil P model, shown as nonnormalized regression coefficients of a path analysis in Ultisol A horizons (% figures indicate the partial correlation between connected fractions. Variability in labile resin P_i is explained to 80% by variations in bicarbonate P_o fraction). Source: Tiessen et al. (1984a).

regions concentrated on methods that assess mainly P_i forms (Olsen and Sommers, 1982; Maida, 1978). Due to the increasing cost of fertilizer P and use of conservation tillage practices, with a consequent build up of organic matter and P in surface soil, increased attention is being given to the importance of P_o as a source of plant-available P. For example, several studies have attributed the lack of a crop response to fertilizer P several studies have attributed the lack of a crop response to fertilizer P application (Doerge and Gardner, 1978; Lathwell, 1979; Reddy, 1983) and seasonal available P variation (Dormaar, 1972; Jessop et al., 1977; Nguyen et al., 1969; Weaver and Forcella, 1979; Stewart and McKercher, 1983) to P_o mineralization. It has also been shown that applied fertilizer P can be incorporated into the P_o pool. For example, in pasture systems Blair et al. (1976) reported that 28 and 40% of applied-fertilizer P (18 kg P ha^{-1} per year) was measured as P_o, 7 and 28 days after application and Rixon (1966) found that 82 to 100% (174 kg P ha^{-1} per year applied) was transformed into P_o.

There are two main approaches for studying the plant availability or "bioavailability" of P_o:

1. Sequence-type studies (Jenny, 1980) where changes in P_o over different soil types, climosequences, or management systems are examined to see what transformations occur and which P_o forms are available.
2. Turnover studies where residues are added to soil and P uptake by plants or forms in the soil are measured to characterize the transformations and availability. This type of incorporation can be in the field or laboratory. In the field, this can be following the fate of added residues under fallow or crop (Dalal, 1982, Dormaar, 1972).

For sequence-type studies, P_o has been measured either by fractionation (Friend and Birch, 1960; Agboola and Oko, 1976; Tiessen et al., 1983; O'Halloran et al., 1985) or by measuring specific compounds using nuclear magnetic resonance (Newman and Tate, 1980; Tate and Newman, 1982; Hawkes et al., 1984). In many of these studies, it has been observed that P_i levels in cultivated soils (where some P is removed by harvest) are the same as uncultivated soils whereas P_o levels decline indicating P_o utilization. Also, differences in bicarbonate ($NaHCO_3$)-extractable P_o values between cropped and fallow years, point to the contribution made by this fairly labile fraction of soil organic matter to crop growth (Fig. 6–4). This figure also illustrates the importance of the silt and clay size fractions on the physical stabilization of root exudates and microbial products.

For turnover studies, P_o has been measured using fractionation procedures (Adeptu and Corey, 1976; Abbott, 1978; Chauhan et al., 1979, 1981; Bowman and Cole, 1978b; White and Ayoub, 1983; Enwezor, 1976; Hedley et al., 1982; Elliott et al., 1984), measuring specific compounds (Lau et al., 1982; Newman and Tate, 1980), and use of tracers (P-32 or P-33) (Harrison, 1982). In tracer P incubation-type studies, the amount of P uptake from the labelled residue can also be determined (Fuller et al., 1956; Till and Blair, 1978; Blair and Boland; Dalal, 1979).

Limited information is available, however, on the relative importance of the mineralization of P_o forms to phosphate ions in solution as sources of P available for plant uptake. One study (Sharpley, 1985) investigated the cycling of P in unfertilized and fertilized agricultural soils

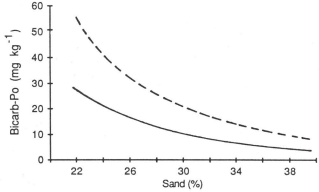

Fig. 6–4. Predicted and observed bicarbonate extractable organic P (bicarb-P_o) levels with increasing sand content in a Mollisol. Source: O'Halloran et al. (1985). The dashed line (---) represents the regression line of soils sampled at the end of the second wheat crop of a fallow-wheat rotation and the solid line is the regression line of soils sampled at the end of the fallow year on the same plots. The difference between values represent potentially mineralizable P_o.

during a 2-yr period under natural field conditions in the Southern Plains area of Oklahoma and Texas to evaluate the relative importance of P_i and P_o as sources of plant-available P. It was found that the Bray-I content of P in a surface silt loam soil (0–50 mm) at El Reno, OK was closely related to the P_i in fertilized soils and to P_o contents in the unfertilized soils (Fig. 6–5). These results demonstrate the respective importance of P_i and P_o as sources of plant-available P in fertilized and unfertilized soils.

A more quantitative estimate of the potential contribution of P_o to plant-available P was determined by calculating the net mineralization of P_o during the period of maximum crop growth, as the decrease in P_o during spring and early summer (Table 6–1). Similar amounts of available P were added to the surface soil from P_o mineralization as from fertilizer P, although a slight decrease in mineralization was apparent in the fertilized compared to unfertilized soils. Even so, the amount of P_o mineralized was related to the total P_o content of each soil (Fig. 6–6) for Houston Black clay at Riesel, TX. The proportion of P_o mineralized annually in the fertilized soils, was significantly lower (at the 0.1% level) than that in unfertilized soil (Table 6–1).

The mineralization of P_o as reported by several other workers is summarized in Table 6–2. A general grouping of slightly weathered soils from temperature regions and more weathered soils from tropical areas has been made. For the temperate soils, no distinct difference in amount of P_o mineralized was apparent between land use types (Table 6–2). This is consistent with investigations of P cycling in forest, grassland, and agricultural ecosystems in the northeastern USA (Grove, 1983). He found that although the mineralization rate was similar in native (4.6 kg P ha^{-1} per year) and agricultural systems (5.1 kg P ha^{-1} per year), the relative

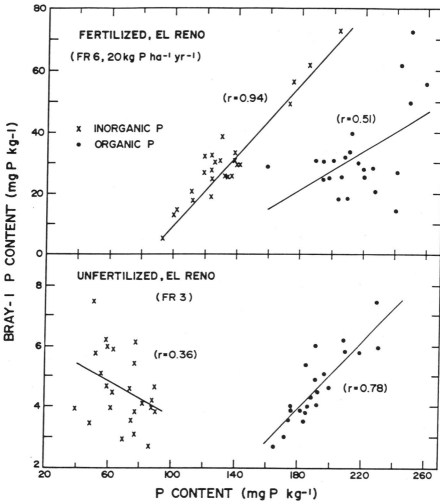

Fig. 6–5. Relationship between plant-available P (Bray-1 P) and inorganic and organic P content of surface Kirkland silt loam (0–50 mm), Riesel, TX for 1981 and 1982. Adapted from Sharpley (1985).

fraction of P uptake supplied by P_o mineralization was lower in cultivated (31%) than native ecosystems (83%) due to a higher P uptake in the former.

The absolute and relative amounts of P_o mineralized are, in general, greater for the more weathered soils from the tropics than for temperate soils (Table 6–2). This is consistent with the fact that areas of the tropics having higher soil temperatures and distinct wet and dry seasons, have more favorable conditions for rapid mineralization of P_o compared to temperate regions. In both tropical and temperate regions, the decom-

Table 6-1. Fertilizer P applied and organic phosphorus mineralized during 1981 and 1982.

Site	Soil	Crop	Fertilizer phosphorus applied	Organic phosphorus mineralized†	Percent mineralized‡
			— kg P ha⁻¹ per year —		%
El Reno, OK					
FR1	Kirkland silt loam	Native grass	—	29	19
FR2		Native grass	—	21	15
FR3		Native grass	—	17	17
FR4		Native grass	—	13	16
FR5		Wheat	20	14	10
FR6		Wheat	20	15	13
FR7		Wheat	20	17	11
FR8		Wheat	20	15	12
Riesel, TX					
Y	Houston Black clay	Mixed	38	19	19
Y2		Mixed	30	21	19
Y6		Cotton/oat sorghum rotation	24	15	18
Y8			19	10	14
Y10			28	20	18
Y14		Klein grass	—	27	20
W10		Coastal Bermudagrass	—	35	22
SW11		Wintergreen Hardinggrass	—	37	26
Woodward, OK					
WW1	Woodward loam	Native grass	—	28	25
WW2		Native grass	—	14	25
WW3		Wheat	25	15	21
WW4		Wheat	25	21	22

†Represented by the decrease in soil organic phosphorus content during February to June 1981 and 1982.
‡Percent of total organic phosphorus content (prior to spring mineralization) mineralized each year of study.

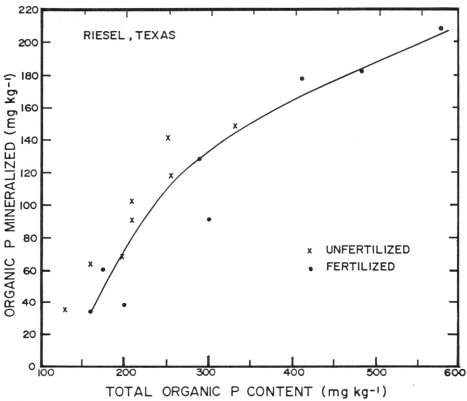

Fig. 6–6. Net mineralization of organic phosphorus as a function of total organic phosphorus content of surface Houston Black clay (0–50 mm), Riesel, TX for 1981 and 1982. Adapted from Sharpley (1985).

position of organic matter and consequent P_o mineralization, is rapid following clearing or cultivation of pristine land. In the tropics, however, this initial mineralization is much greater, with values as high as 50 to 60% mineralization in 3 yr being reported (Brams, 1971; Cunningham, 1963). This releases a large amount of P_i, usually more than a crop can remove. Even so, P deficiencies often develop within a few cropping seasons, due to rapid chemisorption of mineralized P_o (Adepetu and Corey, 1977). Of particular importance to the weathered tropical soils is that lime application, which is often required to overcome Al toxicity and increase crop yields, can stimulate P_o mineralization (Awan, 1964; Dalal, 1977; Thompson et al., 1954). Lathwell (1979) reported that liming representative Oxisols and Ultisols from Puerto Rico, Brazil, Ghana, and Peru increased the availability of soil P_i for several crops, as a result of more favorable conditions for microbial activity increasing P_o mineralization.

Table 6–2. Mineralization of organic phosphorus as a potential source of plant-available P.

Location	Land use	Soil	Study period	Organic phosphorus mineralized	Percent organic phosphorus mineralized/year	References
			yr	kg P ha⁻¹ yr	% yr⁻¹	
Slightly weathered, temperate soils						
Australia	Grass	—	4	6	4	Blair et al., 1976
Australia	Wheat	—	55	0.3	0.3	Williams and Lipsett, 1961
Canada	Wheat	Silt loam	90	7	0.4	Tiessen et al., 1982
		Sandy loam	65	5	0.3	
England	Grassland	Silt and sandy loam	1	7–40	1.3–4.4	Brookes et al., 1984
	Arable	Silt and sandy loam	1	2–11	0.5–1.7	
	Woodland	Silt loam	1	22	2.8	
England	Cereal crop			0.5–8.5		Chater and Mattingly, 1980
England	Deciduous forest	Brown earth	1	9	1.2	Harrison, 1978
England	Grass	Brown earth	1	14	1.0	
Iowa	Row crops	Clay loam	80	9	0.7	Sharpley and Smith, 1983
Maine	Potatoes	Silt loam	50	6	0.9	
Minnesota	Alfalfa	Silty clay loam	60	12	1.2	
Mississippi	Cotton	Silt loam	60	5	1.0	
	Soybean	Silty clay loam	40	8	1.0	
New Mexico	Row crops	Loam	30	2	0.4	
Texas	Sorghum	Clay	60	7	1.0	
Weathered, tropical soils						
Honduras	Corn	Clay	2	6–27	5.9–11.9	Awan, 1964
		Clay	2	10–22	6.9–8.8	
Nigeria	Bush	Sandy loam	1	123	24	Adepetu and Corey, 1977
	Cocoa	Sandy loam	1	91	28	
Ghana						Cunningham, 1963
—Cleared Shaded	Ochrosol fine	3	141	6		
—Tropical half shaded	Sandy loam	3	336	17		
—Rainforest exposed		3	396	17		

Management Considerations

It is clear from studies of the cycling of P in soils in the Southern Plains and other areas, that while P_i is a major source of available P in fertilized soils, P_o can be an important source in both fertilized and un-fertilized soils. Consequently, agricultural management practices which maintain or increase soil P_o contents can be advantageous for long-term soil fertility maintenance. With the increased use of reduced tillage prac-tices and incorporation of plant residues in the surface soil, a build up of soil organic matter will occur. This may not, however, result in a direct increase in P_o content. Chauhan et al. (1981) have shown that cellulose addition to surface soil (Ap horizon) increased microbial activity and, subsequently, the amount of microbial P_o and P_i immobilized. In a P-deficient subsurface soil (Bm horizon), however, an increase in P_o was only observed when cellulose plus fertilizer P was added. Consequently, the application of fertilizer P to reduced-tillage practices, which conserve soil organic matter, may not only increase available P_i content directly, but may also stimulate P_o production. This process will be of particular importance in soils of high P fixation, which maintain available P at low levels. Under these conditions, therefore, fertilizer P may increase both immediate and long-term soil fertility through respective increases in P_i and P_o content.

Another important consideration of P cycling in reduced-tillage prac-tices is that a smaller proportion of applied-fertilizer P is converted to unavailable forms, due to its concentration in the surface soil, compared to conventionally tilled soils (Kunishi et al., 1982). This pool of plant-available P, however, is not in close proximity to the zone of active root growth. The management problem, thus, is how to get this available P as well as mineralized and mineralizable P_o from the surface soil to the zone of active root growth. Mechanical incorporation of residues in the soil may be necessary after several years for more efficient nutrient re-cycling. It is apparent, therefore, that there is a need for research on the cycling of P in soils under these management practices, investigating the effect of factors such as climate, fertilizer P application, and residue type and age on the rate and amount of P immobilized and mineralized.

It is clear that P_o can be an important source of plant-available P in not only tropical but temperate soils and in soils under reduced-tillage practices. For example, significant correlations between P_o and P uptake were found by Acquaye (1963) for cocoa (*Theobroma cacao* L.) in Ghana, by Friend and Birch (1960) for wheat (*Triticum aestivum* L.) in East Africa, and by Omotoso (1971) for cocoa in Nigeria. Consequently, it appears that for soil P fertility tests in both temperate and tropical areas, prediction of P availability might be improved by accounting for the readily mineralizable P_o content as well as P_i. Abbott (1978), Adepetu and Corey (1976), Bowman and Cole (1978a), and Daughtrey et al., (1973) found that the actual potential soil P_i supply as reflected by crop yields, was more closely estimated by including extractable P_o with P_i. Bowman and Cole (1978a) used a modification of the Olsen bicarbonate test (Olsen et al., 1954) which measured the total amount of P (inorganic plus organic) extracted by the reagent. Where other soil P test methods are recommended, a similar adaptation may be used. A potential problem in estimating extractable P_o as part of a soil P test method, is that the conditions of P extraction may not represent conditions for P_o mineralization in the field. In addition, seasonable variations in climatic and soil conditions will have a greater effect on the amount of extractable P_o than P_i content of soil (Dormaar, 1972). Consequently, caution must be exercised in relating amounts of extractable P_o to expected crop response in the field. For instance, O'Halloran et al. (1985) compared the amount of $NaHCO_3$-extractable P_i and P_o in soils in the fallow year of a fallow-wheat rotation against values obtained at the end of the second wheat crop (Fig. 6–4). Clear differences were found in $NaHCO_3$-P_o values with the wheat soil samples much higher than the fallow soil. Conversion of this P_o to P_i must occur during the fallow period. It is possible that soil P test methods including P_o may have to be calibrated on a more site-specific basis with input of cropping history to supplement information needed by present methods.

SULFUR

Conceptual Sulfur Cycle

Figure 6–7 presents a conceptual S cycle in a manner similar to that discussed for P. The conceptual flow diagram describes the main form of S in soil, the main pathways of transformation and sets the boundaries to the cycle in the grassland soil-plant system under study. Atmospheric and groundwater processes are not documented, except for the net inflow and outflow of various S compounds at the soil-air and soil-water interface. In Fig. 6–7, both soil inorganic (S_i) and organic sulfur (S_o) are divided into labile and stable forms. The diagram is a concept or hypothesis of the known types of transformations that take place in soils and does not attempt to show the quantities or dynamics involved.

The mineralization of S_o represents the dominant source of S in most aerated grassland surface soils as S_o accounts for approximately 90% of total S (Bettany et al., 1973; Freney and Williams, 1983). The mechanisms involved in the mineralization of S_o are largely unknown. Breakdown of organic substrate is mediated by microbial populations and, thus, the transfer of S_o from the various organic pools (labile, clay-protected, and stable S_o) to S_i would be via an intermediate microbial S pool or directly by exoenzymes excreted by microorganisms. The excess amounts of S are released to the soil available pool as inorganic sulfate and the remaining S would be incorporated into the microbial S pool and recycled into the various soil organic fractions.

McGill and Cole (1981) proposed that the mechanisms stabilizing organic carbon, nitrogen, sulfur, and phosphorus in soils are not neces-

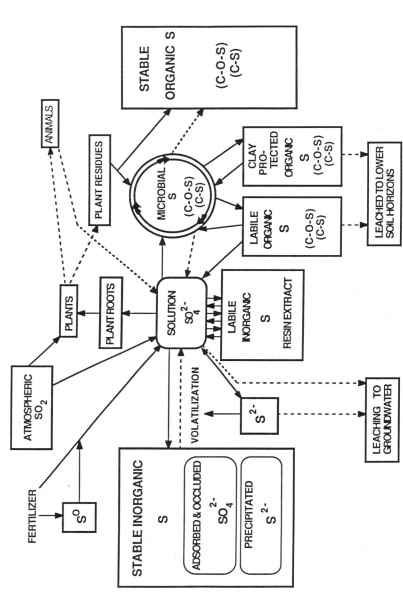

Fig. 6–7. Conceptual flow diagram of the main forms and transformations of S in the soil-plant system. Both soil inorganic and organic sulfur are divided into labile and stable forms. C-O-S refers to ester sulfates and C-S refers to C-bonded S. Source: Maynard et al. (1984).

sarily common to all four elements. The authors suggested that the mechanisms and pathways of S mineralization are specific to the form of S_o. Carbon-bonded S would be mineralized as a result of C oxidation to provide energy (biological mineralization). The need for C not S would result in the mineralization of S. Ester sulfates would be hydrolysized by extracellular or periplasmic sulfohydrolases (produced by soil microorganisms and plant roots) controlled by the end-product supply (biochemical mineralization). The S associated with the ester sulfates would be mineralized in response to a need for S not C. The net mineralization of S would be the result of the combination of the two mechanisms.

Separate controls may also regulate the stabilization or immobilization of the various S_o forms. Sulfate ester production may be a mechanism used by soil microorganisms to store S without altering the pH of their surroundings (McGill and Cole, 1981; Fitzgerald, 1978).

The conceptual model of McGill and Cole (1981) proposed that C and S are stabilized together and mineralized by microbes to provide energy (biological mineralization) and introduced the concept of biochemical mineralization as separate from biological mineralization and controlled by the need for S rather than for C. The concept of stabilization (immobilization) of ester forms independently of, but not necessarily separate from, C-bonded forms was also presented. To these concepts, a third process must be added, that is the eluviation of low-molecular-weight S_o compounds, either with or without clay, which may occur in more weathered soils, thus, causing changes in the composition of S_o with depth (Roberts and Bettany, 1985). Several detailed studies on the processes involved in S mineralization and immobilization using tracers (S^{35}) have substantiated these concepts (Maynard et al., 1983, 1984, 1985). Detailed information on the inputs and outputs of S to soils has been reviewed elsewhere (Freney and Williams, 1983; Bettany and Stewart, 1983). We must now examine how the knowledge gained from conceptualizing and researching processes may be used to explain the supply of S to plants and to improve fertilizer-use efficiency.

Sulfur Transformations During Pedogenesis

Soil S is derived mainly from the weathering of plutonic rocks. As primary minerals weathered, sulfides were released and oxidized to sulfate. This sulfate is either incorporated into soil as organic forms by plants or organisms and as inorganic forms (relatively insoluble salts in semiarid climates) or washed out to oceans via rivers (cf. Freney and Williams, 1983). Soil S, therefore, exists in a wide variety of forms and oxidation states. Within the soil-plant system it is well recognized that most S is in organic combination with several notable exceptions such as saline and acid sulfate soils. Generally, the inorganic pool of soluble and adsorbed sulfates (in well-drained soils) and sulfides (in poorly drained soils) accounts for < 10% of the total S (Bettany and Stewart, 1983; Freney and Williams, 1983). Current analytical techniques have prevented detailed information being gathered on the 90% and more of total S in organic form. Organic sulfur is determined by differences (total S—inorganic S) and S in organic matter has been subdivided into C-bonded S and organic sulfates (linked to C through O and N atom). Early authors commented on the similarity of C/N/S ratios in organic matter of a wide variety of world soils (Whitehead, 1964). More recently, Biederbeck (1978) and Bettany et al. (1973) have stressed the differences in C/N/S ratios and the percentage of S existing as C-S and C-O-S forms that occur as a result of pedogenic processes. Significant changes in the composition and quantity of S_o were found to occur with climate and vegetation across narrow environmental gradients (Bettany and Stewart, 1983). Further work across the same environmental gradient (Roberts and Bettany, 1985; Roberts et al., 1985) showed the influence of topography, as well as climate and vegetation, on C, N, P, and S content of soil organic matter. Other process studies (Maynard et al., 1985) have attempted to relate plant availability of S to composition of soil organic matter in particular the C/S ratio and the percentage of total S occurring as C-O-S. These studies have progressed to the point where computer simulation of processes, have accurately depicted the transformation of S in soil and subsequent plant availability of soil S (Stewart et al., 1984; Hunt et al., 1983, 1986). The conditions simulated by the model are somewhat artificial in that C sources were added evenly throughout the soil. Future work has to bridge the transition to a wider range of management conditions. Eventually, soil testing laboratories need to use past cropping histories and organic matter inputs to predict S fertilizer needs and plant responses to S.

Sulfur-deficiency diagnosis by soil testing methods (Table 6–3) have involved the extraction of soluble sulfate salts with water, a weak electrolyte (0.01 M $CaCl_2$), or phosphate solutions (Hamm et al., 1973; Maynard et al., 1982; Walker and Doornenbal, 1972). These extractions, however, do not account for organic sulfur mineralization. Consequently, deficient soil S levels may be indicated, although no crop response to S

Table 6–3. Some soil extractants of plant-available S (Maynard et al., 1982).

Group 1:	Removes readily soluble sulfate.
1.	Cold water extract
2.	0.15% $CaCl_2$ (0.014 M)
3.	0.1 M LiCl
Group 2:	Removes readily soluble sulfate plus portions of adsorbed sulfate.
4.	400 mg L^{-1} P as KH_2PO_4 (0.016 M)
5.	500 mg L^{-1} P as Ca (H_2PO_4) 2 (0.008 M)
6.	0.03 M $NaH_2PO_4 \cdot 2H_2O$ in N N acetic acid
Group 3:	Removes readily soluble sulfate and adsorbed sulfate plus portions of organic sulfur.
7.	0.5 M $NaHCO_3$ at pH 8.5
8.	Heat soluble S extracted with 1% NaCl
	S extracted with 1% NaCl after moistening the soil and then drying on a water bath and finally in an oven at 100°C

fertilization is obtained. Research is needed, therefore, to develop a soil test procedure which removes the most labile S_o fraction. As was the case for P, however, conditions of chemical extraction are unlikely to simulate mineralization of S_o under field conditions. Nonlabile S_o may also slowly release large amounts of S_i relative to a small labile S_o pool. Improved soil S tests will probably involve the prediction of plant-available S dynamics and turnover, under a wide range of environmental conditions using computer modeling. Refinements in methodology will also be needed before the role of S_o and S_i in plant nutrition can be more comprehensively understood and reliable soil test procedures developed.

CONCLUSIONS

The limitations of present soil test methods for P and S are that they only measure inorganic forms in the soil and do not properly account for the contribution from mineralization of organic forms and past management histories. Inclusion of the contribution of organic matter mineralization has to take into account the seasonality of P and S turnover. Much of the basic information on turnover has not been collected in soil fertility research to date. This is especially important with the more widespread adoption of soil conservation oriented management practices. The challenge to soil science is to develop a complete understanding of P and S cycles in the soil for the development of more firmly based predictive relationships. This will involve both chemical extraction procedures and predictive modelling of organic matter turnover.

REFERENCES

Abbott, J.L. 1978. Importance of the organic phosphorus fraction in extracts of calcareous soils. Soil Sci. Soc. Am. J. 42:81–85.

Acquaye, D.K. 1963. Some significance of organic phosphorus mineralization in the phosphorus nutrition of cocoa in Ghana. Plant Soil 19:65–80.

Adepetu, J.A., and R.B. Corey. 1976. Organic phosphorus as a predictor of plant available phosphorus in soils of Southern Nigeria. Soil Sci. 122:159–164.

––––, and ––––. 1977. Changes in N and P fractions in Iwo soils from Nigeria under intensive cultivation. Plant Soil 46:309–316.

Agboola, A.A., and B. Oko. 1976. An attempt to evaluate plant available P in Western Nigerian soils under shifting cultivation. Agron. J. 68:798–801.

Awan, A.B. 1964. Effect of lime on availability of phosphorus in Zamorano soils. Soil Sci. Soc. Am. Proc. 28:672–673.

Barber, S.A. 1984. Soil nutrient bioavailability. Wiley-Interscience, John Wiley and Sons, New York.

Bettany, J.R., and J.W.B. Stewart. 1983. Sulfur cycling in soils. p. 767–785. In A.I. More (ed.) Proc. Int. Sulphur '82 Conf., Vol. 2, London. 1982. British Sulphur Corp., London.

––––, ––––, and E.H. Halstead. 1973. Sulfur fractions and carbon, nitrogen and sulfur relationships in grasslands, forest and associated transitional soils. Soil Sci. Soc. Am. Proc. 37:915–918.

Biederbeck, V.O. 1978. Soil organic sulfur and soil fertility. p. 273–280. In M. Schnitzer and S.U. Khan (ed.) Soil organic matter. Elsevier Press, Amsterdam.

Blair, G.J., and O.W. Boland. 1978. The release of phosphorus from plant material added to soil. Aust. J. Soil Res. 16:101–111.

----, A.R. Till, and R.C.G. Smith. 1976. Phosphorus cycle—what are the sensitive areas? p. 47–51. *In* G.J. Blair (ed.) Prospects for improving efficiency of phosphorus utilization. Proc. of Symp., Univ. of New England Armidale, New South Wales, Australia. August.

Bowman, R.A., and C.V. Cole. 1978a. Transformations of organic phosphorus substances in soils as evaluated by $NaHCO_3$ extraction. Soil Sci. 125:49–53.

----, and ----. 1978b. An exploratory method for fractionation of organic phosphorus from grassland soils. Soil Sci. 125:95–101.

Brams, E.A. 1971. Continuous cultivation of West African soils: Organic matter diminution and effects of applied lime and phosphorus. Plant Soil 35:401–411.

Brookes, P.C., D.A. Powlson, and D.S. Jenkinson. 1984. Phosphorus in the soil microbial biomass. Soil Biol. Biochem. 16:169–175.

Chater, M., and G.E.G. Mattingly. 1980. Changes in organic phosphorus contents of soils from long-continued experiments at Rothamsted and Saxmundham. Rothamsted Exp. Stn. Report for 1979, Part 2: 41–61.

Chauhan, B.S., J.W.B. Stewart, and E.A. Paul. 1979. Effect of carbon additions on soil labile inorganic, organic and microbially held phosphate. Can. J. Soil Sci. 59:387–396.

----, ----, and ----. 1981. Effect of labile inorganic phosphate status and organic carbon additions on the microbial uptake of phosphorus in soils. Can. J. Soil Sci. 61:373–385.

Cole, C.V., E.T. Elliot, H.W. Hunt, and D.C. Coleman. 1978. Tropic interactions in soils as they affect energy and nutrient dynamics. V. Phosphorus transformations. Microb. Ecol. 4:381–387.

----, G.S. Innis, and J.W.B. Stewart. 1977. Simulation of phosphorus cycling in semiarid grasslands. Ecology 58:1–15.

Coleman, D.C., C.P.P. Reid, and C.V. Cole. 1983. Biological strategies of nutrient cycling in soil systems. Adv. Ecol. Res. 13:1–56.

Cunningham, R.K. 1963. The effect of clearing a tropical forest soil. J. Soil Sci. 14:334–345.

Dalal, R.C. 1977. Soil organic phosphorus. Adv. Agron. 29:83–117.

----. 1979. Mineralization of carbon and phosphorus from carbon-14 and phosphorus-32 labelled plant material added to soil. Soil Sci. Soc. Am. J. 43:913–916.

----. 1982. Effect of plant growth and addition of plant residues on the phosphatase activity in soil. Plant Soil 66:265–269.

Daughtrey, Z.W., J.W. Gilliam, and E.J. Kamprath. 1973. Soil test parameters for assessing plant-available P of acid organic soils. Soil Sci. 115:438–446.

Doerge, T., and E.H. Gardner. 1978. Soil testing for available P in southwest Oregon. p. 143–152. *In* Proc. 29th Annual Northwest Fertilizer Conference, Beaverton, OR. July.

Dormaar, J.F. 1972. Seasonal pattern of soil organic phosphorus. Can. J. Soil Sci. 52:107–112.

Elliott, E.T., K. Horton, J.C. More, D.C. Coleman, and C.V. Cole. 1984. Mineralization dynamics in fallow dryland wheat plots, Colorado. Plant Soil 76:149–155.

Enwezor, W.O. 1976. The mineralization of N and P in organic materials of varying C:N and C:P ratios. Plant Soil 44:237–240.

Field, M.T., and H.F. Birch. 1960. Phosphate responses in relation to soil tests and organic phosphorus. J. Agric. Sci. 54:341–347.

Fitzgerald, J.W. 1978. Naturally occurring organo sulfur compounds in soils. p. 391–443. *In* J.O. Nriagu (ed.) Sulfur in the environment. Part II. Ecological impacts. John Wiley and Sons, New York.

Freney, J.R., and C.H. Williams. 1983. The sulfur cycle in soil. p. 129–201. *In* M.V. Ivanov and J.R. Freney (ed.) The global biogeochemical sulphur cycle. SCOPE Rep. 19. John Wiley and Sons, New York.

Friend, M.T., and H.F. Birch. 1960. Phosphate responses in relation to soil tests and organic phosphorus. J. Agric. Sci. 54:341–347.

Fuller, W.H., D.R. Nielsen, and R.W. Miller. 1956. Some factors influencing the utilization of phosphorus from crop residues. Soil Sci. Soc. Am. J. 20:218–224.

Gebhardt, M.R., T.C. Daniel, E.E. Schweizer, and R.R. Allmaras. 1985. Conservation tillage. Science 230:625–630.

Grove, T.L. 1983. Phosphorus cycles in forests, grasslands, and agricultural ecosystems. Ph.D. thesis. Cornell Univ., Ithaca, NY (Diss. Abstr. 44 (9-2683-B).

Haas, H.J., D.L. Grunes, and G.A. Reichman. 1961. Phosphorus changes in Great Plains soils as influenced by cropping and manure applications. Soil Sci. Soc. Am. Proc. 25:214–218.

Hamm, J.W., J.R. Bettany, and E.H. Halstead. 1973. A soil test for sulfur and interpretive criteria for Saskatchewan. Commun. Soil Sci. Plant Anal. 4:219–231.

Harrison, A.F. 1978. The phosphorus cycles of forest and plant grassland ecosystems and some affects of land management practices. p. 175–195. *In* Phosphorus in the environment: Its chemistry and biochemistry, CIBA Foundation Symposium 57. Elsevier/North Holland, Amsterdam.

––––. 1982. P-32 method to compare rates of mineralization of labile organic phosphorus in Woodland soils. Soil Biol. Biochem. 15:93–99.

Hawkes, G.E., D.S. Powlson, E.W. Randall, and K.R. Tate. 1984. A ^{31}P nuclear magnetic resonance study of the phosphorus species in alkali extracts of soils from long-term field experiments. J. Soil Sci. 35:33–45.

Hedley, M.J., and J.W.B. Stewart. 1982. Method to measure microbial phosphate in soils. Soil Biol. Biochem. 14:377–385.

––––, ––––, and B.S. Chauhan. 1982. Changes in inorganic and organic soil phosphorus fractions induced by cultivation practices and by laboratory incubations. Soil Sci. Soc. Am. J. 46:970–976.

Hunt, H.W., J.W.B. Stewart, and C.V. Cole. 1983. A conceptualised simulation model of C,N,S, and P interactions. p. 303–325. *In* B. Bolin and R.B. Cook (ed.) The major biogeochemical cycles and their interactions. SCOPE 21. John Wiley and Sons, New York.

––––, ––––, and ––––. 1986. Concepts of sulfur, carbon and nitrogen transformations in soil: Evaluation by simulation modelling. Biogeochemistry 2:163–177.

Jenny, H. 1980. The soil resource (Ecological Studies Vol. 37) Springer-Verlag New York, New York.

Jessop, R.S., B. Palmer, F.V. McClelland, and R. Jardine. 1977. Within-season variability of bicarbonate phosphorus in wheat soils. Aust. J. Soil Res. 15:167–170.

Kunishi, H.M., V.A. Bandel, and F.R. Mulford. 1982. Measurement of available soil phosphorus under conventional and no-till management. Commun. Soil Sci. Plant Anal. 13:607–618.

Lathwell, D.J. (ed.) 1979. Phosphorus response on Oxisols and Ultisols. Cornell Int. Agric. Bull. 33. New York State College of Agriculture and Life Sciences, Ithaca.

Lau, C.H., N.K. Soong, and K.S. Tan. 1982. Phosphorus associated with fulvic and humic acids in soils under different covers. p. 37–46. *In* E. Puspharajah and S.H.A. Hamid (ed.) Phosphorus and potassium in the tropics. Malawi Soil Science Society, Kuala Lumpar.

Lindsay, W.L., and P.L.G. Vlek. 1977. Phosphate minerals. p. 639–672. *In* J.B. Dixon and S.D. Weeds (ed.) Minerals in soil environments. Soil Science Society of America, Madison, WI.

Maida, J.H.A. 1978. Phosphate availability indices related to phosphate fractions in selected Malawi soils. J. Sci. Food. Agric. 29:423–428.

Marshall, K.C. 1971. Sorptive interactions between soil particles and microoganisms. p. 409–445. *In* A.D. McLaren and J. Skujins (ed.) Soil biochemistry, Vol. 2. Marcel Dekker, New York.

Maynard, D.G., J.W.B. Stewart, and J.R. Bettany. 1982. Diagnosis of sulfur deficiency in plants and soils. p. 207–228. *In* Proc. Alberta Soil Science Workshop, Edmonton. 23-24 February. Agriculture Soil and Feed Testing Laboratory, Edmonton, Alberta, Canada.

––––, ––––, and ––––. 1983. Sulfur and nitrogen mineralization in soils compared using two incubation techniques. Soil Biol. Biochem. 15:251–256.

––––, ––––, and ––––. 1984. Sulfur cycling in grassland and parkland soils. Biogeochemistry 1:97–111.

––––, ––––, and ––––. 1985. The effects of plants on soil sulfur transformations. Soil Biol. Biochem. 17:127–134.

McGill, W.B., and C.V. Cole. 1981. Comparative aspects of C, N, S, and P cycling through soil organic matter during pedogenesis. Geoderma 26:267–286.

Newman, R.H., and K.R. Tate. 1980. Soil phosphorus characterization by ^{31}P nuclear magnetic resonance. Commun. Soil Sci. Plant Anal. 11:835–842.

Nguyen, Kha, J.C. Vedy, and Ph. Duchaufour. 1969. Etude expérimentale de l'évolution saisonnière des composés humiques en climat tempere. Pedologie 19:5–22.

O'Halloran, I.P., R.G. Kachanoski, and J.W.B. Stewart. 1985. Spatial variability of soil phosphorus as influenced by soil texture and management. Can. J. Soil Sci. 65:475–487.

Olsen, S.R., C.V. Cole, F.S. Watanabe, and L.A. Dean. 1954. Estimation of available phosphorus in soils by extraction with sodium bicarbonate. USDA Circ. 939. U.S. Government Printing Office, Washington, DC.

————, and L.E. Sommers. 1982. Phosphorus. In A.L. Page et al. (ed.) Methods of soil analysis. Part 2. Agronomy 9 (2nd ed.): 403–487.

Omotoso, T.I. 1971. Organic phosphorus contents of some cocoa-growing soils of Southern Nigeria. Soil Sci. 122:195–199.

Reddy, K.R. 1983. Soluble phosphorus release from organic soils. Agric. Ecosystems Environ. 9:373–382.

Rixon, A.J. 1966. Soil fertility changes in a red-brown earth under irrigated pastures. II. Changes in phosphorus. Aust. J. Agric. Res. 17:317–325.

Roberts, T.L., and J.R. Bettany. 1985. The influence of topography on the nature and distribution of soil sulfur across a narrow environmental gradient. Can. J. Soil Sci. 65:419–434.

————, J.W.B. Stewart, and J.R. Bettany. 1985. The influence of topography on the nature and distribution of organic and inorganic soil phosphorus across a narrow environmental gradient. Can. J. Soil Sci. 65:651–665.

Russell, E.W. 1973. Soil conditions and plant growth. 10th ed. Longman Group, Harpow, UK.

Sanchez, P.A., D.E. Bandy, J.H. Villadica, and J.J. Nicholaides. 1982. Amazon basin soils: Management for continuous corn production. Science 216:821–827.

Sharpley, A.N. 1985. Phosphorus cycling in unfertilized and fertilized agricultural soils. Soil Sci. Am. J. 49:905–911.

————, and S.J. Smith. 1983. Distribution of phosphorus forms in virgin and cultivated soils and potential erosion losses. Soil Sci. Soc. Am. J. 47:581–586.

————, and ————. 1985. Fractionation of inorganic and organic phosphorus in virgin and cultivated soils. Soil Sci. Soc. Am. J. 49:127–130.

Smeck, N.E. 1973. Phosphorus: An indicator of pedogenic weathering processes. Soil Sci. 115:199–206.

————. 1985. Phosphorus dynamics in soils and landscape. Geoderma 36:185–199.

Stewart, J.W.B. 1984. Interrelation of carbon, nitrogen, sulfur and phosphorus during decomposition processes in soil. p. 442–446. In M.J. Klug and C.A. Reddy (ed.) Current perspectives in microbial ecology. American Society for Microbiology, Washington, DC.

————, H.W. Hunt, and J.R. Bettany. 1984. Recent progress in understanding the sulphur cycle in soils. p. 695–702. In J.A. Terry (ed.) Proc. of sulphur-84. Sulphur Development Institute of Canada, Calgary, Alberta.

————, D.C. Maynard, and C.V. Cole. 1983. Interaction of biogeochemical cycles in grassland ecosystems. p. 247–269. In B. Bolin and R.B. Cook (ed.) The major biogeochemical cycles and their interactions. SCOPE 21. John Wiley and Sons, New York.

————, and R.B. McKercher. 1982. Phosphorus cycle. p. 221–238. In R.G. Burns and J.H. Slater (ed.) Experimental microbial ecology. Blackwell Scientific Publications, Oxford.

————, and ————. 1983. Phosphorus cycling in soils: Agronomic considerations. p. 501–565. In Third International Congress on Phosphorus Compounds (IMPHOS), Brussels, 4-6 October. Institut Mondial du Phosphate, Casablanca, Maroc.

Tate, K.R. 1984. The biological transformation of P in soil. Plant Soil 76:245–256.

————, and R.H. Newman. 1982. Phosphorus fractions of a climosequence of soils in New Zealand tussock grassland. Soil Biol. Biochem. 14:191–196.

Thompson, L.M., C.A. Black, and J.H. Zoellner. 1954. Occurrence and mineralization of organic phosphorus in soils, with particular reference to associations with carbon, nitrogen, and pH. Soil Sci. 77:185–196.

Tiessen, H., and J.W.B. Stewart. 1983. Particle size fractions and their use in studies of soil organic matter: II. Cultivation effects on organic matter composition in size fractions. Soil Sci. Soc. Am. J. 47:509–514.

————, and ————. 1985. The biogeochemistry of soil phosphorus. p. 463–472. In D.E. Caldwell et al. (ed.) Planetary ecology. Van Nostrand and Reinhold, New York.

————, ————, and J.R. Bettany. 1982. Cultivation effects on the amounts and concentration of C, N and P in grassland soils. Agron. J. 74:831–835.

————, ————, and C.V. Cole. 1984a. Pathways of P transformations in soils of differing pedogenesis. Soil Sci. Soc. Am. J. 48:853-858.

————, ————, and H.W. Hunt. 1984b. Concepts of soil organic matter transformations in relation to organo-mineral particle size fractions. Plant Soil 76:287–295.

----, ----, and J.O. Moir. 1983. Changes in organic and inorganic P composition of two grassland soils and their particle size fractions during 60–90 years of cultivation. J. Soil Sci. 34:815–823.

Till, A.R., and G.J. Blair. 1978. The utilization by grass of sulfur and phosphorus from clover litter. Aust. J. Agric. Res. 29:235–242.

Tisdall, J.M., and J.M. Oades. 1982. Organic matter and water-stable aggregates in soils. J. Soil Sci. 33:141–163.

Walker, D.R., and G. Doornenbal. 1972. Soil sulfate: II. As an index of the sulfate available to legumes. Can. J. Soil Sci. 52:261–266.

Walker, T.W., and J.K. Syers. 1976. The fate of phosphorus during pedogenesis. Geoderma 15:1–19.

Weaver, T., and F. Forcella. 1979. Seasonal variation in soil nutrients under the Rocky Mountain vegetation types. Soil Sci. Soc. Am. J. 43:589–593.

White, R.E., and A.T. Ayoub. 1983. Decomposition of plant residues of variable C/P ratio and the effect on soil phosphate availability. Plant Soil 74:163–173.

Whitehead, D.C. 1964. Soil and plant-nutrition aspects of the sulphur cycle. Soils Fert. 17:1–8.

Williams, C.H., and J. Lipsett. 1961. Fertility changes in soils cultivated for wheat in southern New South Wales. Agric. Res. 12:612–619.

Williams, J.D.H., J.K. Syers, R.F. Harris, and D.E. Armstrong. 1971. Fractionation of inorganic phosphate in calcareous lake sediments. Soil Sci. Soc. Am. Proc. 35:250–255.

7 Current and Projected Technical Support Systems for Producers in Competitive Agriculture[1]

R. A. Olson and J. D. Beaton[2]

The farmer would have to be more than a jack-of-all-trades if he were to make all the decisions required of him in daily operations based on his personal knowledge. He would need to be a trained soil scientist, crop scientist, chemist, entomologist, plant pathologist, agricultural engineer, economist, and animal scientist. The training in each of these disciplines would have to be at or near the Ph.D. level if he were indeed to understand completely the basis for formulating and making those decisions. Quite obviously, a lifetime of study would be required in reaching such a level of training with no time left for being a farmer. Modern agriculture has become so complex in all of its ramifications and so competitive in maintaining solvency as to allow no significant errors in management decisions. Therewith comes the need for reliable information sources for guiding the farmer around the many pitfalls that exist in the soil-plant-animal continuum. The purpose of this chapter is to outline existing support systems in the agronomy area, noting their strengths and weaknesses, and suggesting measures needed in the future.

Most of the discussion will focus on the USA since the authors have had long involvement in technical support systems operating in this country. They recognize, however, that comparable support systems exist in many other developed nations.

TECHNOLOGY SUPPORT ORGANIZATIONS

Government Agencies

U.S. Department of Agriculture

The establishment of the USDA by Congress in 1862 was the first serious step made toward the founding of a scientific agriculture in this

---

[1] Contribution of the University of Nebraska.

[2] Professor of agronomy, University of Nebraska, Lincoln, NE 68583; and Northwest Director, Potash and Phosphate Institute of Canada.

Copyright © 1987 Soil Science Society of America and American Society of Agronomy, 677 S. Segoe Rd., Madison, WI 53711, USA. *Soil Fertility and Organic Matter as Critical Components of Production Systems,* SSSA Spec. Pub. no. 19.

country. Present-day Agricultural Research Service (ARS) has evolved, following several reorganizations in the department from early days, having the specific objectives of planning and conducting research for producing new knowledge and technologies in support of the nation's food and agriculture enterprise. It focuses on long-range problems of national concern that are appropriate for federal research with an integrated and coordinated approach across the entire nation to those problems. The SCS and the Agricultural Stabilization and Conservation Service are sister agencies of the ARS in USDA that work closely with the farmer in implementing erosion control measures and federal agricultural programs. No research is done but substantial technical support is conveyed to the individual farmer in carrying out their mandates.

In June 1886, the Canadian Parliament authorized the Experimental Farm Service, roughly equivalent to the USDA. Approval was given at that time to establish the first five experimental farms in what was a developing and sparsely populated nation. This service was the forerunner of today's Research Branch of Agriculture Canada.

Tennessee Valley Authority

The Tennessee Valley Authority (TVA), thanks to spadework of Gifford Pinchot and Senator George Norris, was created with charge of "planning for the proper use, conservation, and development of the natural resources of the Tennessee River drainage basin and its adjoining territory for the general, social, and economic welfare of the nation. . ." and signed into law by President Roosevelt on 18 May 1933. Flood control, power generation, and navigation were the most evident transformations that it brought about in its early life, but the gradual upgrading of agriculture in the region became apparent by the 1950s. Today its National Fertilizer Development Center is a national and international source of new fertilizer technology and its test demonstration arm carries that technology to industry and farmers of the country. Research at the Center and in collaboration with experiment station workers in most states of the USA has contributed greatly to increased fertilizer-use efficiency by crops. The short courses, workshops, and publications of the Center are major national information sources for the soil fertility-fertilizer use area.

Land Grant Colleges System

The Land Grant Colleges System was created with the Morrill Act of 1862 that provided for state colleges of agriculture and mechanical arts. Some individual states subsequently set up agricultural experiment stations that led to the Hatch Act of 1888 which designated the establishment of agricultural experiment stations in association with the land grant colleges. An annual grant of $15 000 was provided for the station's support in each state and to this day Hatch funds, grown considerably from that early figure, are a major source of *hard* monies for underwriting agricultural research. All of these colleges/universities now have their Agronomy or Crops and Soils Departments that conduct research on the soils and crops common to the particular state. The Extension Service foundation came with the Morrill Act, and was augmented by the Smith-Lever Act of 1914 which set up the three-way teamwork of county, state, and nation in training and providing capable extension leaders.

The several disciplines of crop breeding, crop management, soil fertility and plant nutrition, soil management, entomology, plant pathology, horticulture, agricultural economics, home economics, and agricultural engineering are served by specialists in most states under the experiment station and extension programs. The combination of these programs has resulted in U.S. farmers being the best informed among all the world's farmers. The outstanding progress of the farmers of western Europe in greatly increasing yields of small grains by adopting intensive crop management practices is noteworthy.

Agribusiness

The manufacturers of equipment and chemicals for agriculture and the commercial producers of crop seeds with their advertising are a significant source of information to the farmer. Such releases extolling the qualities of a given product have built-in biases, but do nonetheless supply substantial food for thought to the reader. Modern truth-in-advertising regulations and the compelling need for repeat business make likely the dissemination of reasonably factual information.

Industry associations and institutes complement and expand the capabilities of the individual company in effecting technology transfer throughout the agricultural community. These organizations have their own experts in individual fields quite capable of extending the findings in research and practice of the federal and state support organizations.

The industry institutes also aid in the generation of new research information by encouraging researchers and providing financial support for research of mutual interest. For example, the Potash and Phosphate Institute activity promotes and assists research on maximum yields and maximum economic yield crop production systems through implementation of the "team or multidisciplinary approach" which is necessary to obtain breakthroughs in the complex field of agronomic research.

When all is said and done, the local dealer has the greatest impact on the farmer in his purchase of agricultural inputs. The dealer who keeps himself properly informed has at hand the publications dealing with the specific products he has for sale. The person is the home-town resident, trusted by the neighbor purchaser, familiar with local growing conditions and the customers' management capabilities, and, being the last one in the chain before the product is purchased, has the last and most convincing word.

Private Industry

Commercial soil and plant analysis laboratories started to appear on the scene in the early 1960s as the Chemical Age in agriculture began to take hold. Their role has grown to the point where a majority of soil and plant samples are processed by them and a minority by government-run laboratories that initiated the process (1.5×10^6 government handled samples and 1.534×10^6 commercial in 1979). This is entirely proper since the total number of samples now processed greatly exceeds the state labs' capacity to handle. Frequently, the commercial labs or the dealers they serve provide support staff for collecting samples, saving the farmer the time and effort required for this tedious operation. Care needs to be exercised in the commercial operation that recommendations provided are not excessive in amounts of materials so as to promote greater dealer sales of products than needed that will beget more soil/plant samples.

Some agricultural lenders also provide technical support to their customers. For example, Mr. M. Thornton of the People's Bank of Bloomington, Bloomington, IL has successfully proven the benefits and economic incentives of intensive cropping practices on the more than 21 460 ha in farms managed by his bank.

A most recent innovation in the advisory chain is that of the independent crop consultants. These individuals scout fields at critical times in crop production for ascertaining water, nutrient, and pest control measures needed. They serve the role of advisor in all aspects of soil and crop management, entomology, pathology, and agricultural engineering, involving a range in expertise that the individual farmer could never acquire. The number of consultants in this capacity is growing rapidly with the ever-increasing competitiveness of modern agriculture.

Future

With current trends in declining farmer numbers and reduced state and federal support of agencies providing agricultural information, a growing dependence on industry-supplied guidance seems likely. Recent rapid growth in role of the independent crop consultant would seem to confirm this premise. Federal and state agencies will likely be more and more confined to basic research on agriculture-related issues with less effort expended on applied research and general information dissemination.

TECHNOLOGY TRANSFER

Major Areas of Technical Support

Major areas on which technical support is needed and supplied to farmers on crop production include: crop cultivar/hybrid, soil fertility needs, soil and crop management practices, water management, and disease and pest control practices. Crop cultivar/hybrid trials supply the crop adaptabilty information to the specific region as is needed for effective crop production. Soil and plant analysis laboratories provide information on the nutrient status of the soil and crop with recommendations for overcoming any shortcomings. There are published guidelines for correcting acid and alkali soil conditions; for conserving moisture, maintaining soil organic matter and controlling erosion by tillage, rotation, and mechanical practices fitting the locale; for irrigation and drainage practices appropriate for the crop and region; for weed, insect, and disease control measures; for such crop management practices as cropping sequence, plant population and row spacing; and for the integration of all environmental interactions, such as moisture storage in soil and snow pack, in the soil-plant continuum. These items of information come from the sources that follow.

Publications

Experiment station and USDA bulletins and journal articles supply the most current research information on the various topics cited above. These are complemented by popularized releases, more understandable by the lay audience, and by extension pamphlets, newsletters, that carry specific recommendations to the farmer based on those research results.

Farm journals, magazines, periodicals carry a large pool of production information from federal, state, and industry sources. The articles contained are written for the specific attention of the farming community and in language readily understood by the farmer. Such journals are often a primary means of maximum farmer contact, for example, the *Nebraska Farmer* is known to be received in more than 90% of the state's farm homes.

The agribusiness complex generates a great deal of literature on product characteristics and the proper use of those products. Fertilizer industry-related institutes publish a variety of technical and semi-technical publications for dealer training and farmer instruction. Correspondingly, the dealer in agricultural products with his advertising and releases of promotional material is a major information source and, moreover, provides the one-on-one contact with the farmer that is dominating in impact over all other sources.

Farmer Field Days and Meetings

The adage of seeing is believing is basis for the field days that experiment stations, extension, and industry extensively employ for observing results from production inputs and practices. The impression the observed results create in the farmer's home environment is indelible. The field day has traditionally been a foremost educational tool for advancing the technology of field crop production. A Montana survey taken at field days held in 1982 indicated that 47% of the growers considered such events to be one of their major sources of information. Farmer meetings otherwise in the home environment, mostly during the winter months when time is more available, have also served a major role in keeping farmers abreast of recent research findings in production agriculture. The most effective meetings are those where question and answer occur between farmer and expert, exposing issues and problems confronting the producer at the grass roots level.

Meetings, field days, and conferences have a significant role as well in the further training of specialists in all disciplines who provide direction for the producer of agricultural products. The interaction by a scientist with others in his subject matter area at national and international levels generates new ideas and enthusiasm for advancing research on and ultimate knowledge of production factors.

Radio, TV, and Videotapes

The airwaves have become a primary means of advertising agricultural products in recent times. Visualizing results in pictorial form as achieved on television has become a most cost-effective procedure for the manufacturer of those products. These media have also been used extensively for educational programs to benefit agricultural producers. Public broadcasting stations are most extensively employed for this purpose with up-to-the-minute reports on weather, markets, and other crop-related issues. The TV broadcasts and videotapes afford eye contact with experts on the various topics, adding credibility for the viewer.

Computer Networks

Farm home use of computers has increased fourfold since 1980. According to a 1983 *Successful Farmer* magazine survey, 5% of U.S. farmers owned a home computer. In 1985, another 24% are estimated to have purchased a home computer with an additional 15% in 1986.

The most recent intervention into the support system has been the interactive computer network with software covering many phases of the production and marketing of agricultural products. A prime example is the AGNET system established at the University of Nebraska in 1975 with staff at Lincoln as well as universities in North Dakota, South Dakota, and Washington State (AGNET, 1984). Today, this self-funded network has clients in 47 states and 10 foreign countries, including producers, governments, educational and financial institutions, and a broad base of agribusinesses. These clients access the system with *dumb* terminals or microcomputers, using state or national WATS lines, direct distance dialing, or the GTE Telenet data packet switching network.

Approximately 200 AGNET programs are available under general groupings of production management, marketing information, financial management, consumer interests, and miscellaneous interests. The management models incorporate the expertise of subject matter specialists at various universities, and as such, are based on proven research. Information programs also contain specialists' analyses, along with USDA reports and press releases, commentaries from private businesses, and state-generated reports on weather, plant diseases, and pest infestations. Specific examples include the following:

The *Flexcrop* program, developed in Montana for the AGNET Computer System, was intended to aid in the replacement of the common crop-fallow method of farming with a more intensive and flexible cropping system. Flexcrop considers a number of important production factors including the amount of plant-available soil water at seeding time; an estimate of growing-season precipitation; and grower management inputs, such as crop to be grown, previous crop, weed problems, levels of available soil N and P, and planting date.

Cropfile is the most recent computer program generated in Montana to assist farmers and other crop managers with crop planning. It is based largely on Flexcrop along with some ideas taken from another AGNET program called *Croppak* which was devised by North Dakota State University.

Micro-Form-U-Share fertilizer dealer support package set up by the TVA for optimization of bulk blends, solutions, and suspensions. It also has capability for applicator calibration, nozzle-size selection and invoice preparation.

The *Irrigate* program developed at the University of Nebraska (Lincoln) for computerized projection of irrigation schedules.

The *Blitecast* program of Pennsylvania State University for outlining control measures for potato late blight.

The University of Nebraska's *Soilloss* program for erosion control.

Montana State University's *Soilsalt* program for diagnosis and reclamation of saline and sodic soils.

Home study courses on timely topics such as soil fertility, weed control, and canola production were successful in the provinces of Alberta and Manitoba. Alberta Agriculture in cooperation with the University of Alberta provides an interesting and effective combination of multi-media instruction. A structured combination of printed correspondence lessons, small and large meetings with specialists reviewing pertinent material, newspaper articles and radio and television presentations is used for courses directed to preregistered farmers.

Manitoba schedules at least one major home study course yearly. Course material is offered primarily in the form of printed correspondence lessons on topics related to soil-crop management.

In Nebraska, "Soil as the Plant Sees It" is a 2- and 3-day school which presents various soil fertility topics in an in-depth manner to farmers, fertilizer dealers, and other interested people in the agriculture community. The school with text for home study is designed to provide the basic information needed about the soil and its ability to supply both native and fertilizer nutrients to crops. The school stresses the efficient use of fertilizers to reduce costs and prevent unnecessary contamination of surface and groundwaters.

Crop Focus, also in Nebraska, is a meeting for farmers designed to cover topics of current interest that stress increasing production efficiency and net profits. Topics include all aspects of crop production from selecting the seed to storage. Each session includes 10 topics. A proceedings is provided for home study. Meetings are held at 10 to 15 different locations throughout the state with a different agent at each location.

Future

Observations of trends from the recent past suggest that the visual media and computer networks will serve expanding roles in technology transfer in the future. No individual is capable of fully understanding all of the complex biological mechanisms involved in modern agriculture and of formulating most effective corrective measures for all problem situations as they arise. The visual elaboration of these mechanisms and situations on the TV screen afford clarity for understanding beyond anything that is put into print or expressed verbally. By integrating all of the known interacting variables the computer intervention assists in decision making based on proven principles and practices, thereby eliminating much of the guesswork of prior times.

EFFECTIVENESS OF TECHNOLOGY TRANSFER

Effective Program Examples

There can be no question about the effectiveness of information dissemination programs worldwide in the period since World War II. The Green Revolution materialized in developing countries during the 1960s as a direct result of crop variety improvement programs of the International Center of Corn and Wheat Improvement (CIMMYT) and International Rice Research Institute (IRRI) international organizations. The package of crop variety and fertilizer, water and pesticide inputs developed by these organizations and carried into the countries by their own staffs in cooperation with local extension and research groups converted a number of food deficit countries to lands of plenty. The subcontinent of Asia is a prime example. Other crop improvement programs carried by various UN action agencies and by U.S. universities, as the International Winter Wheat Performance Nursery (IWWPN) of Nebraska and the Oregon State genetic material distribution program, have contributed significantly to the acceptance and use of higher-yielding seed sources. The FAO recognized fertilizer as the spearhead to agricultural development and through its Fertilizer Programme, entering its 25th yr in 1986, effectively demonstrated what fertilizers could do to increase crop yields in traditional agricultural production systems. The results obtained by this and other international aid programs were responsible for food and fertilizer being virtually synonomous terms at the 1974 World Food Conference in Rome. Finally, the more than doubling of average yields of most crops in the USA since 1950 and the exceptionally high yields of small grains now achieved routinely in western Europe suggests that the word on practices for agricultural enhancement does get around.

Characteristics of Producer Population

Farmer acceptance of technological advances occurs first with the most progressive individuals in any given farming community. Usually, these are also the individuals with the most formal education among their peers and who are reasonably free of financial stress. The combination makes them the more willing to take chances with new technologies. They can be characterized as innovators who are not bound by tradition and are the particular ones at whom research releases and extension programs are aimed. These are the leaders who are followed eventually by their neighbors after the latter observe that the new application does indeed work in the local environment. For the followers, the innovation must be quite convincing to bring about acceptance. They, however, are the large majority and the ones who must eventually be brought around if a country's agriculture is to reach its full potential. Failing to accept for financial or whatever other reasons, they become the dropouts leaving increased acreage to be farmed by the most-efficient producers

Government programs have a significant impact on farmer adoption of new technologies. In developing countries government intervention will normally be required in the form of subsidies or at least special credit schemes that allow purchase of the new input, such as fertilizer or pesticide. Where advanced agricultural production systems already exist, as in the USA and western Europe, and where market prices for crops are depressed from over-production governmental incentives for reduced plantings are often counterproductive. The farmer employs that much more of advanced technologies for raising yield levels on his best land that is farmed with resulting greater total production than existed before the acreage curb.

Factors Retarding Adoption of New Technologies

Financial return is the ultimate basis for acceptance of new or advanced technologies by the producer. The large responses of crops to fertilizer N on lands that had received little or no N supplement other than occasional legume growth and manure application since sod breaking and the associated economic returns account for the spectacular growth of the N fertilizer industry in the USA and world during the past 30 yr. Nitrogen remains a rather inexpensive input in relation to the returns it provides such that growth in its use continues even during a period of depressed prices for agricultural commodities. There is place for growth of other essential elements of plant nutrition like P and K, although less spectacular than N, as demonstrated by maximum economic yield studies around the country and world. In China, the returns from N fertilization have been declining due in part to greater incidence and severity of K deficiency. The current ratio of $N/P_2O_5/K_2O$ consumption in China is 1.0:0.26:0.04 contrasted with the ratio of 1.0:0.5:0.2 recommended by Chinese agricultural authorities.

Actually, it was difficult to drum up much enthusiasm for introduction of new technologies that increased production under the market conditions that existed in the summer and fall of 1985. Economic conditions thus become a dominant reason for slow adoption of advanced practices. If, on the other hand, a new measure will reduce unit production costs its acceptance is rapid even under depressed conditions. One example is the practice of reduced tillage. Research has shown that at least equivalent yields can be obtained compared with conventional tillage, reducing production costs with an added side benefit of assisted erosion control. Projections now indicate that most of the country's arable land will be so treated in the next 10 yr.

A more recently recognized disincentive for at least the expanded use of agricultural chemicals is that of environmental impact. Agriculture is increasingly being placed under the gun in agriculturally advanced countries for its contribution to the pollution of ground- and surface-waters by those chemicals. Most often recognized as a hazard in this respect is the occurrence of nitrate (NO_3^-) in groundwater, the major source of drinking water in the USA. Furthermore, the Environmental Protection Agency (EPA) estimates that 50 of the registered pesticides have the potential for leaching to groundwater (CAST, 1985). At least 12 of these have been detected in groundwater under certain conditions, and some have exceeded acceptable health-advisory concentrations. Eutrophication has become a household word, abhorred because of its spoiling effects on streams and lakes heavily committed to recreational purposes. Further finding of excesses of agricultural chemicals in the food chain, for example, NO_3^- in fresh vegetables, exacerbates the problem.

Future

Existing local surpluses of food grains notwithstanding, there is no question of a vastly expanding need for foodstuffs worldwide as rapidly as world population is growing. If and when political and distribution barriers can be overcome, the agricultural industry would need to make full use of all recent and forthcoming technological advances in meeting that need. There are few remaining easy frontiers into which agriculture can be expanded for providing the extra production. It thereby behooves the producer to utilize those production requisites supplied by industry in the most efficient way possible so as to obtain the maximum increase in productivity and quality of product with least detriment to environmental integrity. This notably expands the research responsibilities of state, federal, and industry scientists beyond the traditional role of learning only how to produce more per unit land area. Let us hope that current austerity measures in governments do not squelch this necessary effort before it gets off the ground.

As Donald A. Holt (1985) stated, "Like the steel and automobile industries, production agriculture is a highly competitive international industry. We must increase the flow of well-developed agricultural technology, including computer technology, to American farmers or we will lose our position of leadership in world agriculture." Included in this expressed concern is the underlying need to slow the soil erosion process that is undergirding future U.S. agricultural productivity. Accelerated erosion proceeds at a faster rate than ever in the past, accentuated by the fence row-to-fence row cultivation advocated in the early 1970s and by the ever-growing number of acres that the individual farmer must tend to remain in business.

SUMMARY AND CONCLUSIONS

Numerous technical support systems are available to assist in management decisions so that producers remain competitive in today's agriculture. As a result, many operations can be completed more efficiently than was ever possible in the past. Virtually every farm household has a radio and television set to provide regularly information programs of state and federal agencies. Industry, too, has a major role in promotional programs for its various products.

Facilities and personnel exist for monitoring essentially all aspects of the production chain. State crop improvement associations assure seed quality augmented by the competition among seed-producing companies in providing the best possible seed for each specific environment. Modern laboratories are readily accessible to evaluate the soil's chemical properties in a given field. Therefore, the prescription of fertilizer and other treatments to maximize the soil's productivity is easily obtained. State and federal entomologists and plant pathologists, recently complemented by the independent consultants, follow the incidence of pests and diseases of major crops throughout the growing season to allow immediate corrective measures if needed. For the irrigator, irrigation scheduling has become a scientifically based operation with atmospheric conditions of temperature and humidity, kind and stage of crop growth, and existing soil moisture status fed into the computer. Further precision is built in with the introduction of rainfall probability statistics for the time interval and specific site under consideration. Telephone hookup with an interactive computer network will provide answers to most of the questions that may arise in current production systems.

A trend is evident of an increasing role of industry specialists in the information dispersal system for agriculture. The Land Grant College system of research and extension releases will certainly continue with likely greater emphasis on basic than applied studies by the research arm. Hopefully, the extension arm and the private consultants who are rapidly becoming a foremost one-on-one contact with farmers will develop a good working relationship.

Increased attention will have to be focused on the environmental impact of any long-term treatment. The 97% of the population that is urban is increasingly insistent that air, water, and food be contaminant-free. The federal EPA, USDA, and USDI have monitoring programs to see that these mandates are met. It is, accordingly, more important than ever that commercial, university, and other advisory groups maintain open channels of communication so that excesses for which agriculture could be blamed are kept in check. How unfortunate for the entire agricultural industry if allowable rates of fertilizer, pesticide, irrigation water, and all other inputs were dictated from Washington.

REFERENCES

AGNET. 1984. The first decade. Central AGNET, University of Nebraska, Lincoln.

Council of Agriculture Science Technology. 1985. Agriculture and groundwater quality. CAST Rep. 103.

Holt, D.A. 1985. Computers in production agriculture. Science 228 (4698): 422–427.

8 Improving Crop Management and Farm Profitability: New Approaches for Advisory Services[1]

R. C. Ward, A. D. Halvorson, and K. Wisiol[2]

Farmers are under severe economic pressure. Their production costs have tended to inflate while commodity prices have decreased. Current farm management emphasizes increased input efficiency. Optimum balance of plant nutrients is essential for obtaining high economic yields. Soil testing and plant analysis are our best sources of information for making accurate fertilizer recommendations. Accurate fertilizer recommendations aid farmers in achieving the required nutrient balance without over- or under investing in fertilizer.

University and federal research agencies have traditionally provided the methodologies and correlation data used in soil testing. Many commercial soil-testing laboratories have adapted methodologies recommended by universities in each state or geographic region. Analytical results from different soil-testing laboratories are often comparable, but fertilizer recommendations among the same laboratories may vary drastically (Olson et al., 1982). This large variation in fertilizer recommendations is causing farmers to become skeptical about the value of soil testing. There is an urgent need to re-establish credibility by standardizing fertilizer recommendations as much as possible. Recommendation uniformity will be improved when all crop-production factors that affect yield potential are considered. Use of crop-production computer models adapted to a given geographic region by crop advisors will help accomplish this goal. This will assure that sufficient but not excessive fertilizer is recommended to meet the yield potential of individual fields, which is influenced by a number of crop management factors.

[1] Contribution from Ward Laboratories, Kearney, NE, and USDA-ARS, Akron, CO, and Urbana, IL.

[2] President, Ward Laboratories, Kearney, NE 68847; soil scientist, USDA-ARS, Akron, CO 80720; and collaborator, Dep. of Agronomy, Univ. of Illinois, Urbana, IL 61801 (formerly plant physiologist, USDA-ARS, Urbana).

Copyright © 1987 Soil Science Society of America and American Society of Agronomy, 677 S. Segoe Rd., Madison, WI 53711, USA. *Soil Fertility and Organic Matter as Critical Components of Production Systems,* SSSA Spec. Pub. no. 19.

CROP MANAGEMENT FACTORS AND THE COMPUTER AGE

The term *crop management* includes a critical group of production factors. Some of the vital crop management inputs are (i) rates of fertilizer and lime; (ii) pesticide applications; (iii) planting date and method; (iv) variety selection; (v) plant population; (vi) residue management and seedbed preparation; (vii) soil water status, irrigation needs, and internal drainage; (viii) cultivation; (ix) harvesting methods; (x) crop rotations; and (xi) storage facilities. Some scientists may not consider some of these inputs to be significant, but to the grower and crop consultant all of these factors are vitally important. A systems approach is needed to unify these factors for farmers.

The availability of microcomputers at all levels of farm management (i.e., advisory services to farm homes) has created a real potential for making important management decision-making information more readily available to the end user. Computers can store and retrieve information with ease. Telecommunications makes it possible to access computer programs and data bases stored on mainframe computers located many miles from the user. Weather records are a good example of this capability. Research, extension, and industry personnel are continuing to develop both simple and complex computer models for microcomputers that can aid in making farm-management decisions. Many of these programs will operate on the personal computers used in many farm homes, the Cooperative Extension Service offices, and the agricultural industry.

Advanced soil testing equipment can output test results directly to computers. Incorporating such capabilities together with fertilizer recommendations into an overall data-handling system (a network) will improve the efficiency of laboratory operations. This computer network can store and retrieve current and past information about a field, and can access additional information (for example, precipitation records) and models needed to make a fertilizer recommendation. Such use of computers helps refine and improve the accuracy of fertilizer recommendations as well as increase the efficiency of laboratory operations. Advisory services are encouraged to implement this capability in the immediate future.

Many soil-testing laboratories now use computers for making fertilizer recommendations; however, few consider all of the crop production factors mentioned above in making those recommendations. If they were to do so, in the near future soil-testing laboratories and advisory services could become the users and guardians of the information, programs, and data essential for operating a profitable farm enterprise.

MANAGING SOIL FERTILITY WITHIN CROPPING SYSTEMS

Crops are known to grow satisfactorily for a given climate when supplied with proper amounts of essential plant nutrients. Farmers are using this information to enhance their productivity. In the USA, the Cooperative Extension Service and others have long encouraged and educated farmers to use commercial fertilizer to increase crop production, quality, and profits. This approach assumed, among other things, that maximum yield had top priority since crop prices would far outrun costs. However, the need to educate farmers to fertilize for optimum economic yields is becoming apparent as the cost-price squeeze intensifies and the whole farm economy worsens.

The profitability of fertilizer application is based on yield response. The immediate need of growers is to know how much fertilizer can profitably be applied. Fertilizer application is not always the answer to increasing farm profits. Other factors have to be considered and measured before an optimum economic yield can be predicted. Crop yield on a given soil is influenced by such factors as climate, crop and variety, pest control, cultural practices, soil moisture relationships, rooting depth, fertility level, and physical and chemical conditions of the root zone. Setting a yield goal which considers these factors is often more reliable than just asking the producer's judgment of yield goal. Unrealistic yield goals can result in excessive (or inadequate) fertilization and possibly contribute to environmental pollution.

For example, lack of weed control results in loss of soil-water intended for the crop. This, in turn, results in reduced yield potential and consequently lower fertilizer requirements. Good weed control is essential for optimizing farm profitability.

Seeding date can also influence yield potential. Generally, spring crops respond more favorably to fertilizer applications if the crop is seeded early (Black and Siddoway, 1977; Christensen, 1975). Late seeding may make it necessary to reduce fertilizer recommendations to maintain profitability.

Growing-season precipitation, evapotranspiration demands, and temperature greatly influence crop-yield potential. These factors must be considered when determining yield potential. Yield potential in turn affects the level of plant nutrients required. Dryland crop yields in the Great Plains are directly correlated with plant-available water (soil water and precipitation) (Black and Ford, 1976; Greb, 1983; Halvorson and Kresge, 1982). Therefore, plant-available water must be considered when making crop management decisions and determining yield potential.

Previous crops can have an effect on water use (Black et al., 1981) or soil-water depletion. Use of a monoculture cropping system can result in reduced yields, particularly when a field is cropped every year (Elliott et al., 1978). This problem is not as great where a summer-fallow period is used between crops. Proper crop rotation also helps reduce plant disease and insect infestations and helps break up specific weed pressures. Again, changes in yield potential due to crop rotations will also result in changes in fertilizer nutrient requirements. Stubble management and tillage can also influence the quantity of soil water stored and available for crop growth.

Legume crops reduce the N fertilizer requirement for any nonlegume crop that follows. Most soil-testing laboratories have legume N credits built into their fertilizer recommendation programs. Research has shown a 7 to 12% yield advantage when maize (*Zea mays* L.) or grain sorghum [*Sorghum bicolor* (L.) Moench] follows a legume crop in the rotation (Voss and Shrader, 1979; Classen, 1982). Advisors can show farmers how rotations increase yield potentials, reduce operating expenses, and increase profitability. Application of manure or other organic wastes and removal or return of crop residues can also affect fertilizer (N, P, and K) requirements.

Soil ecologists and microbiologists (Tu and Trevors, 1985; Coleman et al., 1984; Jansson and Persson, 1982) state that soil organisms play a key role in maintaining and improving soil productivity. Organic carbon from soil organic matter, crop residues, and other organic wastes serve as the energy source for these organisms. The soil organisms mineralize nutrients from organic matter and residues into simple forms useable by plants. They also help plants absorb nutrients, as in the case of mycorrhizae. Soils with the right mix of active soil organism populations have a potential to produce higher yields. Soil microflora and their interactions with soil microfauna influence nutrient availability. Soil fungi, bacteria, nematodes, microarthropods, and earthworms all influence the nutrient cycle. Yield potential can be lowered by detrimental soil organisms, including parasitic nematodes and disease-inducing bacteria and fungi, or detrimental mixes of soil organisms. Crop rotation, soil pH, and soil physical properties can be altered to control activity of soil organisms.

Tillage systems may change crop-nutrient requirements. The first year of a no-till or a reduced-tillage system may require greater inputs of fertilizer N relative to inputs for conventional-tillage systems than for later years. Thus, adjustment in fertilizer recommendations may be needed, particularly if the fertilizer is to be broadcast on the soil surface without incorporation. Banding fertilizer below the most biologically active surface-soil layer (>7-cm deep) may reduce the microbial tie-up of any applied N, but may also increase the potential for leaching losses. This practice could therefore reduce the need for a higher rate of fertilizer N in no-till systems during the initial years of using the system (Hoeft and Randall, 1985).

Soil physical and chemical characteristics are major detriments of potential crop yield and fertilizer needs. Soil texture influences the water-holding capacity of a soil and the amount of soil water available for plant growth. A coarse-textured soil has a lower yield potential under dryland conditions than a silt loam soil which has a greater water-holding capacity. Quantity of soil organic matter influences the amount of N mineralized annually from a given soil. Consequently, soil organic matter content can influence the amount of N fertilizer needed to optimize crop yields. The calcium carbonate ($CaCO_3$) or the Fe and Al ion content of a soil will influence the availability of any fertilizer P applied.

Advisory services, with the aid of computers and models, can assist farmers in working through these complex interactions, translating them into practical management recommendations, as more research makes them more understandable. Expert-system computer programs are being built to further advance capabilities to manage these complex interactions.

Kiniry et al. (1983) and Scrivner (1985, personal communication) have developed a computer model that calculates a soil-productivity index based on several soil parameters, measured in 10-cm increments to a depth of 100 cm. These parameters are pH, bulk density, clay concentration, organic matter content, and available soil P and K values. Using such a computer-estimated productivity value, a crop advisor is able to characterize problem fields. This would help the crop advisor and farmer estimate the effect of treatment on productivity and whether costs of improvement can be justified. In some soils, subsoil P and K are positive factors for increasing soil productivity. Thus, fertilizer rates need to be adjusted for all of these factors.

SOIL AND PLANT ANALYSIS IN FERTILITY MANAGEMENT

Soil testing is the best method available to an advisor for making fertilizer recommendations to a grower. Land grant universities have continued fertilizer rate studies over many years. This research has greatly refined fertilizer recommendations. Recently, McLean et al. (1982) have added more sophistication to soil testing. A quick test method was developed to accurately predict the amount of P or K fertilizer a soil requires for optimum yield potential. In the near future, analytical tests like these will be used to predict fixation properties of soils, so that a crop advisor can recommend appropriate fertilizer placement and more accurately suggest "build-up" for each soil type.

Crop advisors or consultants need to use proper soil-sampling techniques and soil test methods adapted and calibrated for a given geographic area. The advisor needs to thoroughly sample soils, including both the surface and subsoil. A sampling depth of 0 to 15 cm is essential for such elements as N, P, K, secondary and micronutrients, and for pH and soil organic matter. In addition, the 15- to 60-cm and 60- to 120-cm depths must be sampled and analyzed for nitrate (NO_3^-) and S to most accurately estimate available soil N and S. With ridge-till, no-till, and other reduced tillage methods, the advisor will have to learn new methods for soil sampling and proper procedures for interpretation of soil analyses. The U.S. Great Plains and the western states have encouraged subsoil sampling for many years. There is currently an interest in deep sampling in the Corn Belt, northeastern, and southeastern areas of the USA. Sampling time will become more important, especially when mobile nutrients are to be measured. Field size is expanding rapidly, and new approaches to sampling are needed.

Advisors need to consider sampling and making fertilizer recommendations based on soil type. Recently, Hest (1985) and Luellen (1985) reported that computer technology has been teamed with a multi-hopper fertilizer spreader for the purpose of adjusting fertilizer application rate to soil type on-the-go. A soils map is made showing the various soil types across the field. The nutrient needs for each soil type, based on soil test results, are input into the computer. The soils map is displayed on a video screen, with the truck's position represented on the screen by the cursor. Fertilizer rates and formulations are changed automatically as the fertilizer applicator moves across the field, encountering the different soil types and nutrient requirements. This same system is capable of herbicide application based on organic matter, pH, a weed map, and other inputs.

Soil test interpretation is one of the most critical steps in making fertilizer recommendations. Interpretation must be based on correlation data researchers supply. An understanding of the sufficiency concept is necessary to understand the responsiveness of a crop under a given set of conditions. University of Nebraska soil test recommendation research (Olson et al., 1982) has shown that the profitability of a fertilizer application is dependent upon the need for the nutrients applied. Their research pointed out several aspects of fertilizer recommendations that are most significant to crop consultants. These are (i) subsoil nutrients contribute to plant growth; (ii) nutrient application in excess of sufficiency levels will not decrease yields; and (iii) environmental, energy, and economic concerns are also involved in fertilizer recommendations.

Plant analysis is another agronomic tool that crop advisors (agronomists) may use to define fertility needs. Plant analysis has traditionally been used as a monitoring tool for measuring nutrient uptake 1 yr for adjustment the next year. Since crop advisors start scouting fields early in the growing season, many fertility problems can be diagnosed early using plant analysis. With plant analysis, the nutrient deficiency can often be diagnosed and treated within 3 or 4 days. The correction of a fertility problem early in the growing season often increases the potential profitability for the growing crop. Proper sampling is crucial for interpretation. Comparison of normal and abnormal samples is often desirable for interpretation.

Adams et al. (1985) developed a computer program for wheat (*Triticum aestivum* L.) plant analysis that monitors the N, P, and K content of wheat plants for the purpose of fertilizer management. The program gives advice to the producer on when to collect plant samples, when to apply fertilizer, and the application rate. The sampling and recommendation process is repeated up to five times during the crop year, depending upon the producer's enrollment date.

NEW TECHNOLOGY FOR SOIL FERTILITY MANAGEMENT

A new profession has developed since 1975 in the farming community. This profession is *crop consulting*. For example, more than 405 000 ha (1 million acres) of Nebraska crop land received crop consulting advice during the 1985 growing season (Lloyd Andersen, 1985, personal communication). A crop consultant gathers pertinent crop production information about a field and then advises the farmer of the management practices to use for maximum economical production. During the growing season, the consultant visits each field on a weekly or semi-weekly basis to scout for pests and in some cases to schedule irrigation. The recommendations for pest control are based on economic thresholds developed by university and commercial research. Irrigation scheduling is usually based on water depletion in the effective root zone and projected evapotranspiration for a selected time period.

Initially, crop consultants developed a program based on integrated pest management (IPM) research. But as the crop consulting business expanded, consultants found that farmers needed as much advice on agronomics as on pest management. Damage that chewing insects caused concerned farmers, but they were not as atuned to observing the more obscure effects of soil compaction, excess fertilization, or nonuniform plant population. Now farmers demand more complete crop management programs. Thus, crop consulting has developed into a reputable profession that is demanding management information from agronomic scientists.

Since crop advisors oversee the complete management of a field, soil and crop scientists must develop a systems approach to crop production that meets their needs. Currently, the agronomic advisor studies the field, visits with the farmer, takes soil and water samples, and checks rooting depth and compaction. He then advises the producer on all aspects of crop production, such as: (i) which variety or hybrid is best suited for the soil and management; (ii) planting rate; (iii) fertilizer rates; (iv) tillage practices (e.g., deep chiseling or residue management); (v) insecticides; (vi) herbicides; (vii) seed treatment; (viii) planting date; (ix) irrigation scheduling; (x) insect control during the growing season; (xi) harvest date and reduction of harvest losses; and (xii) numerous other factors that affect farm profitability.

Crop advisors can help farmers standardize their records so that information can be shared and more easily analyzed. Farmers may want these advisors to keep their records. Farmers need to become more aware of the potential this site-specific information has for improving both yield and profitability.

Doster et al. (1983) suggested that farmers in the Corn Belt might increase net income $25/ha ($10/acre) by selecting and applying fertilizer, seed, chemicals, and tillage according to soil type. To do this, a farmer must be able to (i) determine soil type as he travels across his fields; (ii) switch rates of chemical application on-the-go; (iii) adjust planting and/ or tillage depth on-the-go; and (iv) collect yield and moisture information from the different soil type locations.

As previously noted, technology is becoming available to customize soil management. Computers will soon routinely be used to collect and store data, to recommend various management practices according to soil and crop needs for part of a field, and guide machinery in varying the soil treatment as needed for each part of the field. An inventory of computer software for farmers or farm advisors to use includes more than 700 agricultural extension programs (Strain and Simmons, 1984). In addition, several hundred companies supply commercial software for farmers. Magazines such as *AgriComp* and *Farm Computer News* have arisen to inform computer-owning farmers about these programs.

Several commercial programs are designed to aid soil management decisions, and a number of university and extension soil management programs are available. Some soil fertility programs are included in farm management packages for use on personal computers or in farm program networks, such as AGNET, that are accessible by personal computers. Few programs, however, include field mapping so that farmers can get customized recommendations for different parts of a field based on soil test results from each part. An example is SOIL PLAN (Wisiol et al., 1985), a program developed by USDA-ARS and extension scientists at the University of Illinois.

SOIL PLAN, a microcomputer program, estimates the amounts of N, P, K, and limestone to apply to a field to meet a given yield goal for seven Corn Belt crops. The program allows farmers to get recommendations for P, K, and limestone that are site-specific within a field. Recommendations are based on soil test results and soil type or characteristics, past crops, and crop management plans. If a current soil test is not available, recommendations are adjusted using cropping history and the record of fertilizer use since the last test. Soil type can be selected from a display list of Midwest soils, or soil characteristics can be entered directly if the type is unknown.

SOIL PLAN fertilizer and limestone recommendations are calculated using soil test results, cropping plans, yield goal, soil characteristics, field history, and the quality of the limestone available. If there are results from several soil sampling sites, up to 11 for a field, these values can be entered on a field map on the display screen. The map is easily tailored to fit the shape of the field. The program outlines reasonable groupings of recommended amounts of fertilizer on the map, if a field is variable enough to call for different rates in different areas, or even no application in some areas. A table is displayed showing amounts added to or subtracted from the general fertilizer recommendation due to past or planned management of this field. The user can easily change certain information and compare effects of changes.

FLEXCROP is a microcomputer program model developed for the dryland farmer in the Great Plains (Halvorson and Kresge, 1982). This program helps farm managers evaluate the effects of their crop and soil management decisions on potential crop yield and to decide whether to crop or summer-fallow a given field. In the model, a percentage yield increase or decrease from a base yield is calculated for each management factor. The factors considered are previous crop, planned crop selection (such as spring or winter wheat, *Triticum aestivum* L.; barley, *Hordeum vulgare* L.; oats, *Avena sativa* L.; or safflower, *Carthamus tinctorius* L.), amount of stored soil water, growing season precipitation, variety selection, planting date, weed competition, soil fertility levels as determined by soil test, and crop nutrient requirements (fertilizer needed) to attain the predicted yield potential.

The FLEXCROP program calculates a yield potential based on the amount of plant-available water that the user entered. The plant-available water includes soil water, which is estimated by knowing soil texture and depth of moist soil, and growing-season precipitation, which the user can input or obtain from long-term weather records stored in the computer data base. This yield potential can be increased if a variety is to be grown that has a higher yield potential than the variety used in calculating the relationship between grain yield and plant-available water. Estimated grain yields are then reduced for late planting, poor crop rotations, poor weed control practices, and inadequate levels of plant nutrients.

The program makes a fertilizer recommendation that is based on soil test information user supplies and on the proposed method of fertilizer application (banded, broadcast, or a combination of both). The user is then asked to input the quantity of fertilizer to be applied. Grain yields are reduced if less than the recommended amount of fertilizer is to be applied. The user can then estimate economic returns based on this yield potential and the prices that are input.

CROPPAK, a modified version of FLEXCROP, adds the feature of precipitation probabilities and asks the user what risk level he wishes to accept in estimating yield potential (Leholm and Vasey, 1983). CROP-PAK then makes a fertilizer recommendation based on yield potential, once other management factors such as planting date, etc., have been considered as was done in FLEXCROP. Another North Dakota program, WHEATPAC, is available for calculating potential and realistic yield goals for spring wheat by regions of the state (E.H. Vasey, 1985, per. com.).

CROPFILE is another version of FLEXCROP designed to serve as a file for storing data for a given field and for calculating yield potentials based on the information used in FLEXCROP (P.O. Kresge, 1985, per. com.). CROPFILE was developed in a spreadsheet format so that "what if" games can be played with the data base.

Numerous other programs, both simple and complicated, are available to assist in making crop management and soil fertility management decisions. Public and private advisory services need to start using the information and ideas from these programs to improve their fertilizer and crop management recommendations. The programs discussed here were used only as examples of how computers can be used in the crop management decision-making process.

Computers can rapidly store and access vast amounts of information. Why not use computers to help access the large volume of stored knowledge about crop management? Sharpley et al. (1985) have shown that soil P levels can be modeled fairly accurately and future soil test levels projected. Why not use such a model to project or predict a soil test P level, if current soil test P levels are not available but older soil test results are? Programs need to be developed that integrate or access numerous other programs that would help provide the information needed to make a more intelligent fertilizer recommendation.

Many farmers will rely on crop advisors and consultants to collect the pertinent crop data for use in computer models such as these. Many farmers will have several options for getting recommendations, but will probably use an agronomic advisor to help in the data handling. The exception will be the farmer who has an agronomic degree or a strong interest in technical aspects of crop production. The role of the Cooperative Extension Service will be important in linking farmers with timely computer recommendations. An increasing number of farmers may be able to use a microcomputer terminal and have access by telephone hookup to the crop advisor's office or soil analysis laboratory to obtain crop management information for a given field. The Cooperative Extension Service will need to play a key role in training crop advisors and consultants if the system is to function.

FUTURE NEEDS

The end result is that tomorrow's farmer expects to have answers to his agronomic questions. The crop consultant will be a key person on his management team. Many farmers will continue looking to crop advisors or consultants for a personal touch to crop production. This has been described as the human element that is part of the farmer's desire to have support in his management decisions. But in any case, either the farmer or the crop consultant will demand computer models and programs that give consistent and accurate recommendations. These individuals are looking to scientists for information.

Agronomic scientists need to supply or provide computer models for farmers to use, crop advisors, and consultants. Since advisors are customers of soil-testing laboratories, it is logical for the laboratory staff to work with the scientists on agronomic information transfer. A clear understanding of crop rotations, residue management, tillage practices, and fertilizer placement and timing enhance analytical results and recommendations.

Researchers need to become more aware of how their results can be molded to fill the voids in our knowledge base. They need to fit their results into current information-transfer models or develop models of their own that will be useful in information transfer by both private and public advisory services. We need to be mindful that the farmer is the end user of most of our agronomic data and information. Farmers are the ones who eventually put their livelihood on the line and apply the technical information supplied to them. Therefore, scientists and advisors need to do the best job possible in being sure the information they provide to the user is sound and meets current needs. The farmer's needs should be one of the main determinants of the work agronomic scientists undertake.

REFERENCES

Adams, D., S.L. Chapman, T.C. Keisling, and T.L. Stewart. 1985. Monitoring wheat for plant N, P, K content and fertilizer management. Univ. of Arkansas Ext. Computer Tech. Bull. 3.

Black, A.L., P.L. Brown, A.D. Halvorson, and F.H. Siddoway. 1981. Dryland cropping strategies for efficient water-use to control saline seeps in the northern Great Plains. Agric. Water Manage. 4:295–311.

----, and R.H. Ford. 1976. Available water and soil fertility relationships for annual cropping system. p. 286–290. *In* Proc. of Regional Saline Seep Control Symposium, Montana State Univ., Coop. Ext. Serv. Bull. 1132.

----, and F.H. Siddoway. 1977. Hard red and durum spring wheat responses to seeding date and NP-fertilization on fallow. Agron. J. 69:885–888.

Christensen, N.W. 1975. Spring grain fertilizer response in Montana as influenced by seeding date. p. 115–124. *In* Proc. of the 26th Annual Northwest Plant Food Assoc. Fert. Conf., Salt Lake City, UT. 15–17 July.

Classen, M.M. 1982. Crop sequence and N effects on corn and grain sorghum. p. 77–79. *In* Report of Prog. 423. Agric. Exp. Stn. Kansas State University, Manhattan.

Coleman, D.C., R.V. Anderson, C.V. Cole, J.F. McClellan, L.E. Woods, J.A. Trofymow, and E.T. Elliott. 1984. Roles of protozoa and nematodes in nutrient cycling. p. 17–28. *In* R.L. Todd and J.E. Giddens (ed.) Microbial-plant interactions. Spec. Pub. 47. American Society of Agronomy, Crop Science Society of America, and Soil Science Society of America, Madison, WI.

Doster, D.H., R.L. Nielson, S.D. Parsons, and T.T. Bauman. 1983. Computer assisted Big Ten crop farming. Mimeo. Presented at 1983 On-Farm Computer Conf. Purdue University, West Lafayette, IN.

Elliott, L.F., T.M. McCalla, and A. Waiss, Jr. 1978. Phytotoxicity associated with residue management. p. 131–146. *In* W.R. Oschwald (ed.) Crop residue management systems. Spec. Pub. 31. American Society of Agronomy, Crop Science Society of America, Soil Science Society of America, Madison, WI.

Greb, B.W. 1983. Water conservation: Central Great Plains. *In* Dryland agriculture. Agronomy 23:57–72.

Halvorson, A.D., and P.O. Kresge. 1982. FLEXCROP: A dryland cropping system model. USDA Agric. Prod. Res. Rep. 180. U.S. Government Printing Office, Washington, DC.

Hest, David. 1985. New fertilizer rig spreads by soil type. The Farmer 103 (15):21–22.

Hoeft, R.G., and G. Randall. 1985. Tillage affects fertility: How to alter one when you change the other. Crops Soils 37(4):12–16.

Jansson, S.L., and J. Persson. 1982. Mineralization and immobilization of soil nitrogen. *In* Nitrogen in agricultural soils. Agronomy 22: 229–252.

Kiniry, L.N., C.L. Scrivner, and N.E. Keener. 1983. A soil productivity index based upon predicted water depletion and root growth. Univ. of Missouri-Columbia Res. Bull. 1051.

Leholm, A.G., and E.H. Vasey. 1983. "CROPPAK": A short run crop enterprise analysis program available on AGNET—A management tool for Agriculture. North Dakota State University Cooperative Extension Service, Fargo.

Luellen, W.R. 1985. Fine-tuned fertility: Tomorrow's technology here today. Crops Soils 38(2):18–22.

McLean, E.O., T.O. Olaya, and S. Mostaghimi. 1982. Improved corrective fertilizer recommendations based on a two-step alternative usage of soil tests: I. Recovery of soil-equilibrated phosphorus. Soil Sci. Soc. Am. J. 46:1193–1197.

Olson, R.A., K.D. Frank, P.H. Grabouski, and G.W. Rehm. 1982. Economic and agronomic impacts of varied philosophies of soil testing. Agron. J. 74:492–499.

Sharpley, A.N., C. Gray, C.A. Jones, and C.V. Cole. 1985. Estimation of phosphorus model parameters from limited soil survey information. p. 160–163. *In* Proc. of the Natural

Resource Modeling Symp., ARS-30. Pingree Park, CO. 16–21 Oct., USDA-ARS, Washington, DC.

Strain, J.R., and S. Simmons. 1984. Updated inventory of computer programs. Univ. of Florida Cooperative Extension Service.

Tu, C.M., and J.T. Trevors. 1985. Invisible farmhands: Soil microbes have a big job to do. Crops Soils 37(5):12–15.

Voss, R.D., and W.D. Shrader. 1979. Crop rotations—effects on yields and response to nitrogen. Iowa State University, Cooperative Extension Service.

Wisiol, K., R.G. Hoeft, and A. Buttitta. 1985. SOIL PLAN: Fertilizer and limestone recommendations. User's guide and diskettes. Univ. of Illinois, Champaign, IlliNet AGRO-B-113.

9 Nutrient and Organic Matter Dynamics as Components of Agricultural Production Systems Models[1]

C. V. Cole, J. Williams, M. Shaffer, and J. Hanson[2]

Agricultural systems, developed for a variety of climate and soil resources, have achieved remarkably high levels of food, feed and fiber production to provide relative security for an ever-increasing world population. With few exceptions, high production levels are achieved by bringing agronomic practices into harmony with natural processes of the biosphere. Increasingly, we are probing the limits of plant and animal productivity.

Global and regional patterns of energy and water distribution are the overriding controls on biological processes essential for agricultural production. Availability of essential nutrient elements imposes the next level of control. Fertilizer use has largely removed or mitigated this limitation. Fertilization levels, however, that are imbalanced with other processes can result in both short- and long-term adverse effects on the environment and soil productivity. The challenge to modern agriculture is to develop optimum balances between inputs and outputs of production systems while also providing a basis for continued increases in productivity with minimum adverse effects on the environment.

The major driving variables, responses of production systems, and interactive feedbacks for agricultural production systems are shown in Fig. 9–1. Systematic study of each of the driving variables and interactive feedbacks is necessary to analyze and predict consequences of management decisions. This is no more than a formal statement of the working concepts that each farm manager has of the opportunities and constraints in his particular farm operation and the management decisions he makes to optimize both the short- and long-term productivity of his system.

[1] Contribution of USDA-ARS, Fort Collins, CO 80523 and USDA-ARS, Temple, TX 76503.

[2] Research soil scientist, USDA-ARS, Fort Collins, CO; research hydrological engineer, USDA-ARS, Temple, TX; research soil scientist, USDA-ARS, Fort Collins, CO; and research range scientist, USDA-ARS, Fort Collins, CO, respectively.

Copyright © 1987 Soil Science Society of America and American Society of Agronomy, 677 S. Segoe Rd., Madison, WI 53711, USA. *Soil Fertility and Organic Matter as Critical Components of Production Systems,* SSSA Spec. Pub. no. 19.

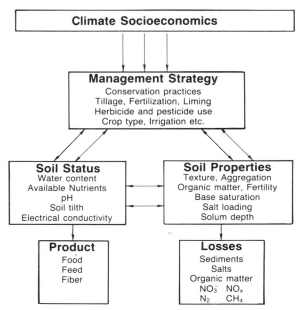

Fig. 9–1. Major controls on productivity of agricultural systems.

Soil properties and soil status in the long and short term, respectively, dictate management strategies within a given climatic and socioeconomic environment. Available water and nutrients, pH, electrical conductivity, and similar properties dictate seasonal adjustments in management decisions. In turn, however, management practices degrade or improve both soil status and properties in a dynamic manner. Harvest of products of the cropping system, the major operational goal, is accompanied by nutrient and crop residue removal. Often other major losses from the system occur that are not associated with harvest. These losses include nutrient leaching, wind- or water-borne sediments, and gaseous effluxes. As a result, off-site impacts such as sediments may have significant environmental consequences as well as modifying the properties of the original soil.

Management practices suitable for rich, fertile soils of the Midwest are obviously inappropriate for semiarid regions. Similarly, deep soils with good water infiltration rates have more productive potential than shallow soils. Topography and nutrient-supplying power are also important constraints. The necessity to return profits to the farm operator is, of course, an overriding control. Although degradation of soil physical conditions and fertility occurs where management practices are out of balance or inappropriate to the resource base, significant improvements in properties can be made by good management practices. This potential for soil improvement is generally underestimated and more attention should be given to positive effects of good management on soil properties.

The need for increased productivity levels, along with stringent economic constraints, are placing heavy burdens on producers. Research information developed for one type of climatic and soil conditions may have limited application under different conditions. The range of management options is wide and varied and is steadily increasing as the result of improved crop cultivars, emerging tillage practices, new opportunities for residue management, additional methods of pest control, and the integration of plant and animal management systems. Developments in biotechnology promise opportunities for improvement of management options. Many costly errors, however, will continue to be made in integrating new management options simply on a "cut-and-try" basis. Most rapid progress will be made by researchers and farm managers who have a good grasp of the overall system and how the different parts interact.

Improved methods are especially needed to interpolate and extend research findings for a variety of conditions, and to transfer these findings to farm managers. Simulation models are being developed with these objectives. These models are tools for organizing information and testing hypotheses about the structure and operation of production systems. Good models are sets of multiple hypotheses about interactions and feedbacks of key processes and their controls. Such models serve to organize and test data sets for consistency, identify gaps in knowledge, and identify relative importance of various processes in their effects on the final product. Models also serve as predictive tools for evaluation of short- and long-term effects of changes in major controls such as climate or different management practices.

Successful modeling requires careful identification of objectives and goals. These will dictate scales of resolution in time and space appropriate for model structure. Simulation models are and will continue to serve as valuable tools for communication among scientists, as vehicles for information transfer, and as decision aids to practical users of management information.

This chapter describes crop production models which integrate organic matter and nutrient dynamics. These models involved over two score scientists representing agricultural engineering, crop physiology, hydrology, range and animal sciences, soil chemistry and microbiology, and soil physics. Previous chapters have documented many processes for the development of sound management practices in crop production systems. With the current trend toward specialized disciplinary backgrounds, interactions and feedbacks of all these processes may be beyond our capabilities and therefore not fully appreciated. Increasingly, this level of understanding requires interdisciplinary team development.

MODELING OF SOIL ORGANIC MATTER DYNAMICS

Abundant information is available in the soils literature on organic matter contents and properties and an increasing understanding of the cycling of C, N, S, and P through organic matter (Stevenson, 1985). Incorporation of concepts of organic matter formation and turnover into simulation models is an excellent means of integrating this information into analyses of agricultural production systems.

The importance of soil organic matter levels to yield potential is highlighted by the relationship shown in Fig. 9–2. Potential maize (*Zea mays* L.) yields under Michigan conditions indicates that the yield potential increases more than 20% for each 1% increase in soil C. As Lucas et al. (1977) documented for Michigan conditions, cropping practices directly affect soil C levels and activity. They developed a model of soil organic matter dynamics which takes account of decomposition, losses through erosion, and inputs from crop residues and manure. Their estimates of annual soil C inputs for several field crops ranged from 369 kg/ha with field bean (*Phaseolus* s.p.) to 632 kg/ha for soybean [*Glycine max* L. (Merr.)] and up to 1243 kg/ha with maize. These inputs are related to yield levels and especially to levels of N fertilization. For a given yield level, each cropping practice has a steady-state soil C level ranging from 0.7 to 1.1% C under Michigan conditions. Common cropping practices require more than 60 yr before a soil approaches a steady-state level. Continued farm manure application will significantly increase the soil C level. High maize yields can more than double soil C content in comparison to low level production.

These relationships are illustrated in Fig. 9–3 which shows the computed annual loss or gain of soil organic carbon as a function of yield levels of maize grain. Note that soil C losses, associated with soil erosion at a rate of 8000 kg/ha, were considered in the calculations. In time, soil C levels approach steady-state levels (zero annual change) for each yield

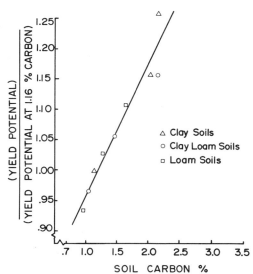

Fig. 9–2. Effects of soil C content on yield potential for corn production in Michigan (Lucas et al., 1977).

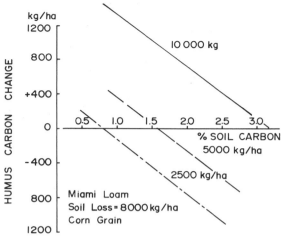

Fig. 9–3. Annual changes in soil humus in relation to maize yield and soil C content of Miami loam (Lucas et al., 1977).

level. This study's significant feature is the strong feedback of soil C effects on maize yield responses to N.

At constant yield levels, other management practices markedly influence steady-state soil C levels as illustrated in Fig. 9–4. When only grain is harvested soil C levels approach 2%, but are substantially lower when silage is removed. Manure additions may more than compensate for silage removal. The study demonstrated strong synergistic relationships between soil C level, N fertilizer efficiency, and maize yields.

Van Veen and Paul (1981) developed a model of organic carbon dynamics to test many of the concepts derived from their review of the literature describing residue decomposition in various parts of the world. Voroney et al. (1981) validated the model and simulated the long-term effects of cultivation and erosion on organic carbon levels of grassland soils. They used the model to test different input conditions, such as rate of plant residue additions and soil erosion, and determine their effects on the equilibrium level of soil organic matter. Their studies showed that erosion had drastic effects on soil organic carbon levels, especially after 50 to 80 yr of cultivation, resulting in a continual decline of organic matter content. (Fig. 9–5).

The Century soil organic matter model (Parton et al., 1983) was developed to predict long-term effects of climate, soil texture, and management on soil organic matter levels and turnover rates. The model was based on concepts of Anderson (1979) and considers, in addition to crop residues, active, slow and passive forms of organic matter as Jenkinson and Rayner (1977) suggested. Crop residues are divided into decomposable and recalcitrant fractions as Paul and Van Veen (1978) proposed. Although the Century model was developed based on information on

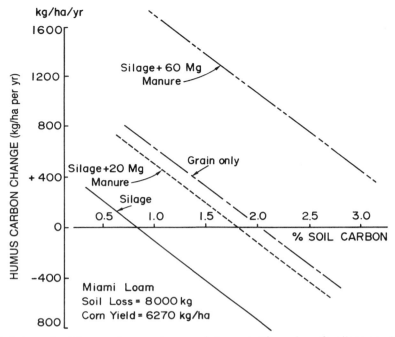

Fig. 9–4. Annual soil humus changes when corn is harvested for grain or for silage, or when manure is applied to corn land harvested for silage (Lucas et al., 1977).

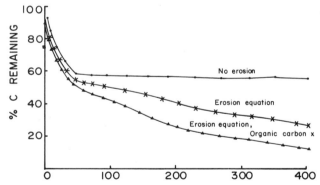

Fig. 9–5. Simulated effects of rainfall erosion on the organic carbon remaining in the surface 15 cm of a cultivated Oxbow soil in wheat crop-crop-fallow rotation. Erosion protection factor 0.75, straw removed in first 50 yr (Van Veen and Paul, 1981).

grassland soils of North America, it satisfactorily simulated long-term changes in Swedish soils (Parton et al., 1982).

   The Century model has been applied to evaluating the effects of erosion on soil organic matter levels with maize-grain production. Table 9–1 shows the simulated affects of erosion on steady-state (long-term)

Table 9-1. Simulated effects of erosion on steady-state levels of soil organic carbon and fertilizer efficiency at two levels of N fertilization (Parton, unpublished).

Nitrogen rate kg/ha	Erosion, Mg/ha			
	0	5	15	30
	Soil C (0.2 m), kg/m^2			
60	8.2	7.7	6.1	3.6
120	11.3	10.5	8.1	5.5
	Fertilizer efficiency, kg grain/kg N			
60	63	59	56	46
120	52	50	45	40

levels of organic carbon under two production levels of maize for grain. In these simulations, C levels following continued application of 120 kg of N/ha with erosion at rates of 15 000 kg/ha are the same as soil C levels at 60 kg of N/ha with no erosion. Changes in fertilizer efficiency expressed as kg of grain harvested/kg of N added are also shown. These results were computed assuming deep soils with no impact of erosion on rooting depth. These results support the approach of Lucas et al. (1977) and indicate the opportunities for examining long-term management practices effects.

MODELING OF NITROGEN TILLAGE AND RESIDUE MANAGEMENT

Model Description

Nitrogen, tillage and residue management (NTRM) (Shaffer and Larson, 1982; Shaffer et al., 1983; Shaffer, 1985) is a simulation model designed to provide research capabilities for studying physical, chemical, and biological processes and their interactions in the soil-crop-water-atmosphere continuum. The model can be used to quickly determine environmental impacts and provide direct management assistance to farmers in an interactive version, COFARM (Shaffer et al., 1984). Nomographs and extension bulletins can be developed from both NTRM and COFARM.

The NTRM (Fig. 9-6) contains integrated submodels for soil temperature, soil C and N transformations, unsaturated water flow, crop and root growth, evaporation and transpiration, tillage, interception and water infiltration, chemical equilibria processes, solute transport, and crop residues.

Decay of crop residues and recycling of C and N are simulated using methods described by Shaffer et al (1983) and Molina et al (1983). Dynamic soil organic matter pools are maintained for surface and incorporated residues at various stages of decomposition and for the soil humus fraction. Process rates involving organic matter and N are simulated

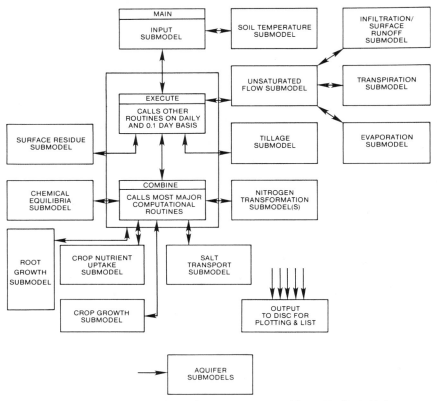

Fig. 9–6. The NTRM simulation model. Adapted from Shaffer (1985).

as functions of appropriate environmental variables including water content, temperature, aeration, pH, and osmotic potential. Plant N uptake combined with root sloughing and decay provide another direct circular pathway between the crop and the soil.

The crop growth submodel consists of a series of integrated subroutines capable of simulating the growth and development of field maize, sweet corn, sorghum (*Sorghum vulgare* L.) soybean, spring and winter wheat (*Triticum aestivum* L.), oat (*Avena fatua* L.), barley (*Hordeum vulgare* L.), rye (*Secale cereale* L.), sunflower (*Helianthus annuus* L.), potato (*Solanum tuberosum* L.), alfalfa (*Medicago sativa* L.), pasture grass, sugarbeet (*Beta vulgaris* L.), cotton (*Gossypium hirsutum* L.), peanut (*Arachis hypogaea* L.), tomato (*Lycopersicon esculentum* Mill.), field pea (*Pisum sativum* L.), sugarcane (*Saccharum officinarum* L.), sweet potato (*Ipomoea batatas* (L.) Lam.), and carrot (*Daucus carota* L.). The NTRM is capable of simulating any combination of these crops in rotational sequence, and several years of simulation can be run to estimate the impacts of crop rotation on various parts of the system.

Calculated values for soil leachate volumes and solute concentrations of chemical ions (Ca^{2+}, K^+, Na^+, Mg^{2+}, NH_4^+, NO_3^-, Cl^-, SO_4^{2-}, HCO_3^-, and CO_3^{2-}) serve as input to a system of submodels for water and solute movement and interactions in the saturated zone (Shaffer et al., 1976). This combination of submodels is particularly useful in studies involving NO_3^--N loading of groundwater as well as irrigation return flow quality and quantity.

The NTRM model inputs include climate data such as daily maximum and minimum air temperatures, pan evaporation, precipitation, wind run, and solar radiation (optional). Values must be provided for soil physical, chemical, and biological properties of user-specified soil horizons. The model contains options which allow estimation of some of these properties by the model in lieu of user-supplied values. The model also requires data concerning management practices such as crop selection and rotational sequences, planting dates, tillage and fertilizer practices, residue management, and irrigation.

Model Examples

Three examples were chosen to demonstrate NTRM's use. The first two examples described demonstrate NTRM's usefulness in simulating crop responses to management and as a decision aid. The second and third examples show the usefulness of NTRM for evaluation of long-term management strategies.

Comparisons of crop yield validation data obtained for the model at Lancaster, WI are shown in Fig. 9–7. These results are typical of model validation obtained at other sites in the northern Corn Belt and demonstrate, by the good agreement with observed values over a wide range of crop yields, that NTRM effectively integrates a wide range of environmental, management, and other input conditions.

Larson et al. (1983) applied the NTRM model to examine various soil management practices proposed for the semiarid Great Plains. A method was developed which allowed a 100-yr frequency for sorghum yields to be calculated using a minimum number of runs with the NTRM model (Fig. 9–8). The results provide the resource manager with an idea of the changes in mean yield and yield distribution for different management practices over the long term.

The NTRM model has been applied to simulate the impact of soil erosion on the short- and long-term productivity of the soil (Shaffer, 1985). Maize yields were simulated, assuming various depths of erosion on a Corn Belt soil. Long- (100 yr) and short-term (10 yr) maize yields were calculated and compared with observed data. In addition, the impact of various N fertilizer and tillage practices on yields were simulated (Fig. 9–9). These studies illustrate how a simulation model can be used to produce information relative to best management practices for various soils and conditions.

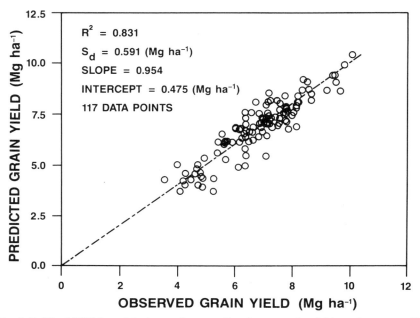

Fig. 9–7. The NTRM model observed vs. predicted corn grain yields at Lancaster, WI [Adapted from Swan et al. (1986)].

MODELING THE IMPACT OF EROSION OF SOIL PRODUCTIVITY

Model Description

Erosion productivity impact calculator (EPIC) was developed to assess the effect of erosion on soil productivity (Williams et al., 1984). The EPIC simulates erosion, plant growth, and related processes as well as economics. Since erosion can be a relatively slow process, EPIC was designed to simulate hundreds of years.

Although EPIC was designed primarily to determine relationships between erosion and soil productivity, it has several other potential uses. The EPIC is capable of assisting with decisions involving drainage, irrigation, water yield, erosion control (wind and water), weather, fertilization, pest control, planting dates, crop varieties, tillage, and crop residue management. Field data are needed for testing and validating the model and, in turn, model results are useful in designing and modifying field experiments to obtain critical information.

The EPIC model contains integrated submodels for hydrology, weather, erosion, nutrients, plant growth, soil temperature, tillage, economics, and plant environment control. The hydrology model simulates the volume of surface runoff water and peak discharge rate given daily rainfall amounts. Other hydrology components include evapotranspira-

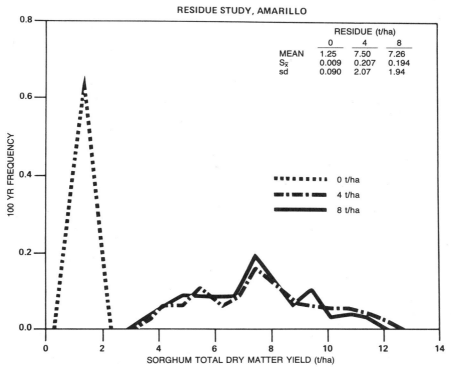

Fig. 9–8. Simulated long-term yield distribution for sorghum, Amarillo, TX. Adapted from Larson et al. (1983).

tion, percolation, lateral subsurface flow, and snow melt. Precipitation, air temperature, and solar radiation information can either be read into the program or they can be simulated. Snowfall is estimated as a function of precipitation and air temperature. The wind (velocity and direction) is simulated if wind erosion is to be considered. The EPIC is capable of simulating both wind and water erosion. The plant nutrients are considered in EPIC are N and P. The N processes that are simulated include runoff of NO_3^--N, organic-N transport by sediment, NO_3^--N leaching, upward NO_3^--N movement by soil water evaporation, denitrification, immobilization, mineralization, crop uptake, rainfall contribution, and fixation.

The N submodel is a modification of PAPRAN (Seligman and van Keulen, 1981). The model considers two sources of mineralized N: fresh organic nitrogen associated with crop residues, and microbial biomass and organic nitrogen associated with the soil humus pool. Nitrogen immobilization is closely linked with residue decomposition and plant uptake from the successive soil layers. The P processes that are simulated include runoff of soluble P, sediment transport of mineral and organic

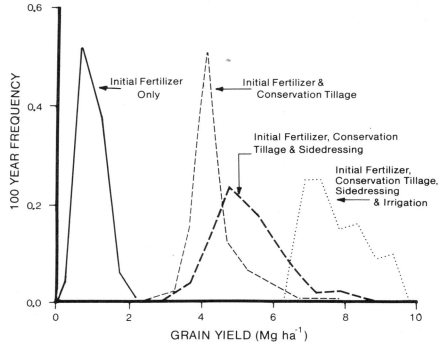

DAKOTA SOIL:

	Mean	S_d
Initial Fertilizer Only	0.91	0.32
Initial Fertilizer & Conservation Tillage	4.45	0.79
Initial Fertilizer, Conservation Tillage & Sidedressing	5.19	0.94
Initial Fertilizer, Conservation Tillage, Sidedressing & Irrigation	7.67	0.80

Fig. 9–9. Long-term distribution of maize grain yields under several levels of management (initial fertilizer is 0.15 Mg/ha N applied at planting). Adapted from Shaffer (1985).

phosphorus, immobilization, mineralization, sorption-desorption, and crop uptake (Jones et al., 1984; Sharpley et al., 1984).

A general plant growth model is used to simulate aboveground plant growth, yield, and root growth for maize, grain sorghum, wheat, barley, oat, peanut, sunflower, soybean, alfalfa, cotton, and grasses. The plant growth model simulates energy interception; energy conversion to roots, aboveground biomass, and grain or fiber production; and water and nutrient uptake. Plant growth is constrained by water, nutrient, and air temperature stresses. Soil temperature is simulated to serve the nutrient cycling and root growth components of EPIC. Soil temperature is predicted at the center of each soil layer as a function of the previous day's soil temperature and the current air temperature, crop residue, and snow cover. The EPIC tillage model simulates ridge height, surface roughness, nutrient and crop residue mixing, change in bulk density, and conversion of residue from standing to flat. The economics component of EPIC uses a crop budget to calculate crop production costs. Income is determined from simulated annual crop yields. The plant environment control component provides options for irrigation, drainage, furrow disking, fertilization, liming, and pesticide application. Complete details of EPIC have been described previously (Williams et al., 1984).

Model Examples

Two examples were chosen to demonstrate EPIC's use in agricultural management. The first example described the development of the erosion-soil productivity relationships for the USA. The second example demonstrates the model's usefulness in a low level management situation in India.

The erosion productivity index (EPI) predicts the sensitivity of soils to erosion in terms of reduced crop production and the potential benefits of improved management practices. Changes in water-holding properties, nutrient supplying capability, organic matter, and soil tilth are prime components needed for prediction of crop yield potential.

Two long-term (100 yr) annual crop yield simulations are required to calculate an EPI for a particular soil. The two runs are designed to represent the entire range of conservation strategies that could be applied to the land. The first run represents ideal conservation totally preventing erosion. Maximum erosion rates, assuming no conservation practices, are simulated in the second run. The two runs are identical (weather, initial conditions, and crop rotation) except for erosion.

As top soil is eroded over subsoils less favorable for root growth, EPI decreases rapidly from its initial value of 1.0. The EPI varies considerably from year to year because of weather and soil water content, especially after extensive erosion. Dry years usually decrease yields more on eroded soils than on uneroded soils because of differences in soil water-holding capacity. Because of the variability in EPI from year to year, a curve-smoothing method was developed to describe the relationship between accumulated EPI and time. To demonstrate the application of the EPI approach, a hypothetical test site was assumed near Lafayette, IN, on a Miami silt loam soil (fine-loamy, mixed mesic Typic Hapludalf). The slope length and steepness were assumed to be 100 m and 5%, respectively. Continuous maize with an annual fertilizer rate of 100 kg ha^{-1} of N and 20 kg ha^{-1} of P was the assumed management strategy. Figure 9–10 shows the progressive effect of erosion of 13 cm of Miami silt loam on the EPI. Soil productivity drops rapidly with time during the early years as the favorable top soil is lost and more slowly as the subsoil approaches the surface.

The EPIC model can be used to estimate the effects of varying one management practice (e.g., irrigation, fertilizer rate, or planting date) while holding all other practices constant. This simulation strategy predicts the

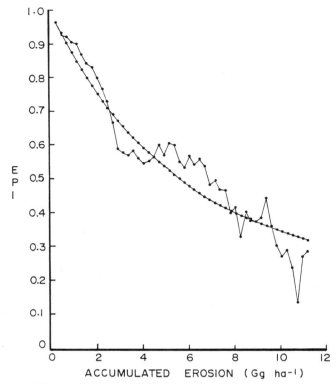

Fig. 9–10. The EPIC simulation of EPI index for Miami silt loam (Williams et al., 1985).

effect of changing a single limiting factor. However, when one factor becomes nonlimiting others may replace it. Thus, entirely new management strategies must often be developed to improve several components of the farming system. This approach was used at the International Cereals Research Institute for the Semi-arid Tropics (ICRISAT) in India for development of new farming systems.

As an example of possible EPIC applications in technology transfer, four management strategies were simulated for the deep Vertisols at ICRISAT and crop growth was simulated over a 50-yr period. The first management strategy is similar to a traditional system. Grain sorghum is planted in the post-rainy season without fertilizer N or P (Table 9–2). In this system, N-deficiency limited mean simulated grain sorghum yield to 547 kg/ha, which is similar to the 483 kg/ha mean yield at ICRISAT from 1976 to 1981 (ICRISAT, 1981). When fertilizer N and P were added to the traditional system, N deficiency was eliminated and simulated grain sorghum yields approximately doubled. Following the elimination of N-deficiencies, cool temperatures and water stress became the most important constraints on grain sorghum yield. To alleviate these stresses, the crop was planted in the rainy season when moisture and temperature

Table 9–2. The EPIC simulation of four different management systems at ICRISAT (50-yr means) (ICRISAT, 1981).

Season/ Crop	Fertilizer		Rainfall	Runoff	Soil loss	Yield	
	N	P				Dry matter	Grain
	kg/ha		——— mm ———		Mg/ha	——— kg/ha ———	
Post-rainy							
Grain sorghum	0	0	809	143	9.1	1450	547
Grain sorghum	45	25	809	101	4.9	3750	1370
Rainy							
Maize	0	0	809	141	6.7	1830	673
Maize	45	25	809	75	2.5	9200	3430

limitations are not as great. In this case, maize was planted instead of grain sorghum because mold-resistant, humidity-tolerant sorghum germplasm is not available at ICRISAT (Kampen, 1982). From 1976 to 1981, the improved maize-chickpea sequential system gave a mean maize yield of 3200 kg/ha (ICRISAT, 1981). The simulated 50-yr mean yield for this system was 3430 kg. Changing to this rainy-season farming strategy and including the use of fertilizer was especially effective for decreasing surface water runoff and erosion, particularly when fertilizer was added (Table 9–2).

MODELING RANGELAND PRODUCTIVITY

Model Description

Simulation of production and utilization of rangelands (SPUR) is a general grassland simulation model for evaluating rangeland systems. The SPUR model provides a basis for management decisions, allows comparison of environmental impacts of alternative management strategies, and forecasts impacts of climatic change on range ecosystems. The SPUR model is composed of five physically based components: climate, hydrology, plant growth, animals (domestic and wildlife), and economics (Fig. 9–11).

Existing weather records of precipitation, maximum and minimum temperatures, solar radiation, and total wind run on a daily basis are used to drive the model. The hydrology component calculates upland surface runoff volumes, peakflow, snowmelt, upland sediment yield, and channel streamflow and sediment. Available soil water used to simulate plant growth is derived using a soil water balance equation. Surface runoff is estimated by the SCS curve number procedure (USDA, 1972) and soil loss is computed by the Modified Universal Soil Loss Equation (USLE) (Williams, 1975). The snowmelt routine employs an empirical relationship between air temperature and energy flux of a dynamic snowpack.

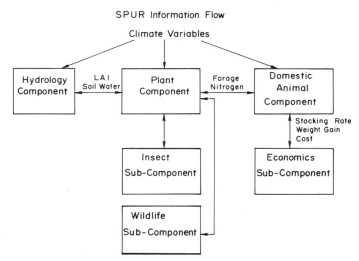

Fig. 9–11. The SPUR model components (Hanson et al., 1982).

Net photosynthesis is used to predict forage production. Carbon and N are cycled through several compartments including standing green, standing dead, live roots, dead roots, seeds, litter, and soil organic matter. Inorganic-N levels with time are also simulated. Required model inputs include the initial biomass content for each compartment and parameters that differentiate the species to be simulated. Temperature, N availability, and soil-water availability control plant processes (Skiles et al., 1982).

The animal component considers domestic livestock and wildlife consumers. Detailed growth information is available for cattle on a steer equivalent basis. Forage consumption is calculated for all classes of animals. The Texas A & M Beef Model (Sanders and Cartwright, 1979) was adapted to compute steer growth. Plant utilization by animals is based on forage palatability, abundance, and location. Wildlife (including insects) is considered as a fixed consumer and is given first access to available forage. Animal production or net gain is used by the economic component to estimate benefits and costs of alternative grazing practices, range improvements, and animal management options.

Two versions of SPUR have been developed. The first version is a grazing unit or pasture-scale version. This version allows grazing animals to differentially graze as many as nine sites and simulates plant growth and grazing responses for a maximum of seven major plant species. It also provides erosion, water runoff, and peak flow indices for relative comparisons of range-site and management option combinations. The second version is a basin-scale version that requires considerable averaging of animal and plant data. The basin-scale version uses the watershed as a management unit. This version is intended to give high resolution for water runoff, peak flow, sediment yield, and channel hydrology. Quantities of runoff and sediment yield for basins with an area of not more than 26 km² and a maximum of 27 hydrologic units can be predicted.

Work is continuing to improve the SPUR's usefulness in the decision-making process. Upon completion, the model will allow one to predict effects of climatic change on rangeland composition, production and stability, effects of grazing system changes or plant environment on cattle efficiency; determining proper stocking rates; and determining the optimum grazing system for a particular parcel of land. The simulation of effects of alternative management strategies on erosion and runoff from various sites is also possible. Such management strategies as grazing systems, stocking rates, tillage, seeding, pest management, water develop-

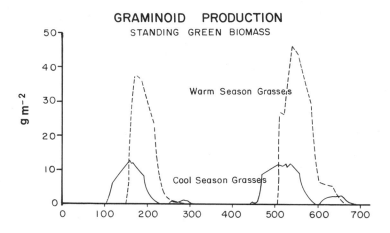

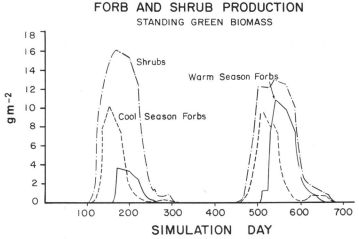

Fig. 9–12. Simulated dynamics of rangeland productivity in eastern Colorado shortgrass steppe.

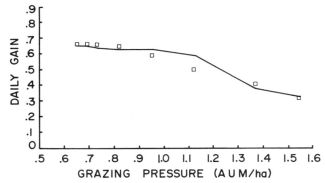

Fig. 9–13. Effect of grazing pressure on animal daily gain (kg day^{-1}) showing SPUR simulated results vs. data of Bennett (1969).

ment, supplemental feed, mixes of agronomic herbivores, irrigation, fire management, and fertilization can be examined.

MODEL EXAMPLES

Rangeland dynamics of the eastern Colorado shortgrass steppe were simulated for 2 yr at moderate stocking rates. Changes in standing green biomass of warm- and cool-season grasses, cold- and warm-season forbs, and shrubs are shown in Fig. 9–12. Changes in soil organic matter and N levels can be examined in a similar manner. These exercises are useful to test for system instabilities requiring adjustments in management.

Simulated effects of grazing pressure on rates of cattle gain are compared with the results of Bement et al.'s (1969) 19-yr study in Fig. 9–13. Although the model tended to slightly overpredict daily gain at intermediate grazing pressures, the adverse effects of overgrazing are illustrated. The results indicate SPUR can be useful to predict animal gains under various range conditions.

SUMMARY

Agricultural systems have achieved high levels of production by bringing agronomic practices into harmony with natural physical, chemical, and biological processes. As the limits of plant and animal productivity are approached, it is vital that information on nutrient and organic matter dynamics be integrated with information on all other factors controlling the functioning of these complex systems. Simulation models are being developed to integrate these sources of information, to interpolate and extend research findings over wide ranges of conditions, and to transfer these findings to farm managers. This chapter describes several crop production models that integrate nutrient and organic matter dynamics into quantitative analyses of system properties and productivity.

Several models have specifically analyzed the formation and turnover of organic matter in relation to inputs of plant residues, decomposition rates, and the concurrent immobilization and mineralization of nutrients. The linear relationship of soil C levels and potential maize yields in Michigan soils was incorporated into a model of organic matter dynamics that demonstrated the strong synergistic relationships between soil C level, N fertilizer efficiency, and maize yields. Other models have been developed to predict long-term effects of climate, soil texture, and management on soil organic matter levels and turnover rates.

The NTRM Model (Shaffer et al., 1982) incorporates nutrient and organic matter dynamics into a comprehensive analysis of the functioning of crop production systems. The model has been applied to a wide range of cropping systems and is serving as a useful tool for researchers. A simplified interactive version is being used to rapidly evaluate environmental impacts and provide direct management assistance to farmers.

The EPIC model was developed to assess the effects of erosion on soil productivity over both short- and long-term timeframes. The ability to predict effects of changing limiting factors is useful in the development of new management strategies to improve components of farming systems.

Nutrients and organic matter are also important components of extensive production systems such as rangelands. The simulation of productivity and utilization of rangelands using SPUR predicts effects of climatic change on rangeland composition, production, and stability as well as the effects of changes in grazing systems or plant environment on cattle gains.

The development of these tools for analysis of a wide variety of agricultural production systems challenges soil scientists to develop and present information on transformations and flows of nutrient and organic matter in a form that may be readily incorporated and utilized.

REFERENCES

Anderson, D.W. 1979. Processes of humus formation and transformation in soils of the Canadian Great Plains. J. Soil Sci. 30:77–84.

Bement, R.E. 1969. A stocking-rate guide for beef production on blue-grama range. J. Range Manage. 22:83–86.

Hanson, J.D., W.J. Parton, and G.S. Innis. 1985. Plant growth and production of grassland ecosystems: A comparison of modelling approaches. Ecol. Model. 29:131–144.

International Crops Research Institute for the Semi-Arid Tropics (ICRISAT). 1981. Improving the management of India's deep black soils. In Proc. Seminar Management of Deep Black Soils for Increased Production of Cereals, Pulses, and Oilseeds, New Delhi, India. 21 May.

Jenkinson, D.W., and J.H. Rayner. 1977. The turnover of soil organic matter in some of the Rothamsted classical experiments. Soil Sci. 123:298–305.

Jones, C.A., C.V. Cole, A.N. Sharpley, and J.R. Williams. 1984. A simplified soil and plant phosphorus model: I. Documentation. Soil Sci. Soc. Am. J. 48:800–805.

Kampen, J. 1982. An approach to improved productivity on deep Vertisols. International Crops Research Institute for the Semi-Arid Tropics, Information Bull. 11. ICRISAT, Patancheru, India.

Larson, W.E., J.B. Swan, and M.J. Shaffer. 1983. Soil management for semi-arid regions. Agric. Water Manage. 7:89–114.

Lucas, R.E., J.B. Holtman, and L.J. Connor. 1977. Soil carbon dynamics and cropping practices in agriculture and energy. p. 333–351. W. Lockeretz (ed.) Agriculture and energy. Academic Press, New York.

Molina, J.A.E., C.E. Clapp, M.J. Shaffer, F.W. Chichester, and W.E. Larson. 1983. NCSOIL, a model of nitrogen and carbon transformations in soil; description, calibration, and behavior. Soil Sci. Soc. Am. J. 47:85–91.

Parton, W.J., D.W. Anderson, C.V. Cole, and J.W.B. Stewart. 1983. Simulation of soil organic matter formation and mineralization in semiarid agroecosystems. R.R. Lowrance et al. (ed.) p. 00–00. In Nutrient cycling in agricultural ecosystems. Spec. Pub. 23. The Univ. of Georgia, College of Agriculture Exp. Stn., Athens.

––––, J. Persson, and D.W. Anderson. 1983. Simulation of organic matter changes in Swedish soils. p. 511–516. In W.K. Lauenroth, et al. (ed.) Analysis of ecological systems: State-of-the-art in Ecological Modeling. Elsevier Scien. Publishing Co., New York.

Paul, E.A., and J.A. Van Veen. 1978. The use of tracers to determine the dynamic nature of organic matter. Trans. Cong. Soil Sci., 11th 3:61–102.

Sanders, J.O., and T.C. Cartwright. 1979. A general cattle production systems model. I. Description of the model. Agric. Systems 4:217–227.

Seligman, N.G., and H. van Keulen. 1981. PAPRAN: A simulation model of annual pasture production limited by rainfall and nitrogen. p. 192–221. In M.J. Frissel and J.A. van Veen (ed.) Simulation of nitrogen behavior of soil-plant systems, Proc. of a workshop, Wageningen. 28 Jan.–1 Feb. 1980. Pudoc, Centre for Agricultural Publication and Documentation, Wageningen.

Shaffer, M.J. 1985. Simulation model for soil erosion-productivity relationships. J. Environ. Qual. 14:144–150.

––––, J.B. Swan, and M.R. Johnson. 1984. Coordinated farm and research management (COFARM) data system for soils and crops. J. Soil Water Conserv. 39:320–324.

––––, S.C. Gupta, D.R. Linden, J.A.E. Molina, C.E. Clapp, and W.E. Larson. 1983. Simulation of nitrogen, tillage and residue management effects on soil fertility. p. 525–544. In Analysis of Ecological Systems: State-of-the-Art in Ecological Modelling. Elsevier Science Publishing Co. New York.

––––, and W.E. Larson (Ed.) 1982. Nitrogen-tillage-residue management (NTRM) model: Technical documentation. Research Report. USDA-ARS, St. Paul.

Sharpley, A.N., C.A. Jones, C. Gray, and C.V. Cole. 1984. A simplified soil and plant phosphorus model: II. Prediction of labile, organic, and sorbed phosphorus. Soil Sci. Soc. Am. J. 48:805–809.

Skiles, J., J. Hanson, and W.J. Parton. 1982. Simulation of above- and below-ground N and C dynamics of Bouteloua gracilis and Agropyron smithii. p. 467–474. In W.K. Lauenroth (ed.) Analysis of ecological systems: State-of-the-Art in Ecological Systems. Elsevier Science Publishing Co., New York.

Stevenson, F.J. 1986. Cycles of soil carbon, nitrogen, phosphorus, sulphur, micronutrients. John Wiley and Sons, New York.

Swan, J.B., M.J. Shaffer, W.H. Paulson, and A.E. Peterson. 1986. Simulating the effects of soil depth and climatic factors on corn yield. Minn. Agric. Exp. Stn. Pub. 15067.

U.S. Department of Agriculture-Soil Conservation Service. 1972. National Engineering Handb., Sec. 4. USDA-SCS, Washington, DC.

Van Veen, J.A., and E.A. Paul. 1981. Organic carbon dynamics in grassland soils. I. Background information and computer simulation. Can. J. Soil Sci. 61:185–201.

Voroney, R.P., J.A. Van Veen, and E.A. Paul. 1981. Organic C dynamics in grassland soils. 2. Model validation and simulation of long-term effects of cultivation and rainfall erosion. Can. J. Soil Sci. 61:211–224.

Williams, J.R. 1975. Sediment-yield prediction with Universal Equation using runoff energy factor. p. 244–252. In Present and perspective technology for predicting sediment yield and sources. USDA-ARS-S-40. U.S. Government Printing Office, Washington, DC.

––––, C.A. Jones, and P.T. Dyke. 1984. A modeling approach to determining the relationship between erosion and soil productivity. Trans. Am. Soc. Agric. Eng. 27(1):129–144.

––––, J.W. Putman, and P.T. Dyke. 1985. Assessing the effect of soil erosion on productivity with EPIC. In Proc. National Symp. Erosion and Soil Productivity, New Orleans, LA. December 1984. American Society of Agricultural Engineers, St. Joseph, MI.